Understanding the Structure of Electricity Supply

Understanding the Structure of Electricity Supply

K.R. Padiyar

BSP **BS Publications**
A unit of **BSP Books Pvt. Ltd.**
4-4-309/316, Giriraj Lane, Sultan Bazar,
Hyderabad - 500 095
Phone : 040 - 23445605, 23445688

Understanding the Structure of Electricity Supply

Published by

BS Publications

A unit of **BSP Books Pvt. Ltd.**

4-4-309/316, Giriraj Lane, Sultan Bazar,
Hyderabad - 500 095
Phone : 040 - 23445605, 23445688
e-mail : info@bspbooks.net
www.bspbooks.net

ISBN: 978-93-85433-87-0 (HB)

Preface

The electrical energy made a beginning about 125 years ago for street lighting. However, due to the tremendous versatility of electricity, and the ability of power engineers to deliver it at affordable costs and high reliability, the electrical energy is regarded as a necessity (along with food, clothing, shelter, education and health care) by citizens all over the world. The electrification was considered the greatest engineering achievement of the 20th century (ahead of the internet).There has been tremendous growth of demand for electric power, which needs to be met by unconventional solutions as the fossil fuels are becoming scarce; the environmental and climate change issues are restricting the choices. The nuclear power does not have the drawback of contributing to carbon foot print, however the safety issues are limiting its application. Large hydroelectric plants have the drawback of environmental problems and displacement of populations. Wind and solar appear to acceptable solutions, but they are difficult to dispatch due to the variability of generation. The energy storage technologies that can help to stabilize the output of renewable generation are not fully mature and need further developments.

The deregulation or restructuring of electric power systems in the recent times has introduced a new paradigm in the operation of ESI (Electricity Supply Industry). The Vertically Integrated Utilities (VIU) have been broken up into three companies-Generation, Transmission and Distribution. Competition has been introduced in power generation and power trading is encouraged to sell the power to the highest bidder in the market through power exchanges. Economists view energy as a product and the transmission lines as highways to route the product. The limitations imposed by current carrying capacity or security constraints are viewed as congestion. The technical problems facing system operation and control are expected to be solved by using Information and Communication Technology (ICT). In India, the Electricity Act of 2003 has resulted in unbundling of generation, transmission and distribution with incentives to private parties to enter generation and transmission sectors. The Electricity Regulatory Commissions (both at center and the states) are established to regulate the ESI. In formulating rules for the system operation, it is necessary to ensure that they are in conformity with the physics of the electricity supply. For example, the Kirchhoff's Voltage Law (KVL) also plays a role in determining the power flows in a line (unlike in transportation networks where only Kirchhoff's Current Law (KCL) is applicable).

This book is primarily directed at non-electrical engineers and even lay persons, to understand the workings of ESI. The only background assumed is high school physics. The major objective is to explain the structure of the existing and future (envisaged) power systems in simple and easy to understand language. The generation, transmission and distribution, system operation and basic controls, protection and system collapse (blackouts), renewable energy sources (RES) and electrical energy storage EES), emerging technologies that can result in smart and self-healing grids are discussed. A proper understanding of the various aspects should guide the system planners, designers, operators and regulators in identifying the problems and seeking optimal solutions.

The book can be used a text or reference for courses on Electricity Supply at any level in colleges where such courses are planned (for students in any branch where energy (and in particular electrical energy) issues are important. Even lay citizens interested in power supply issues can benefit from the book.

- Author

Acknowledgements

I wish to acknowledge my teachers, the authors of books I have studied, Power engineers, my colleagues and students who have contributed directly or indirectly in my understanding of electrical power systems and its operation.

I wish to thank Mr. Anil Shah and Mr. Naresh Daver of the BS Publications for their interest and work towards publication of this book in a record time.

In preparation of this book, Mrs. Sudha Aithal has assisted in the preparation of the manuscript including illustrations. Mr. Saichand has helped in drawing couple of diagrams. Mr. Babu Narayanan has provided data for a typical daily load curve of a distribution company.

Finally, I wish to thank my wife, Usha for her patience and understanding.

- Author

Contents

CHAPTER 3

CHAPTER 4

CHAPTER 5

CHAPTER 1

Why Do Lights Go Out ?

"If anything can wrong, it will."
Murphy's Law

1.1 Introduction

Just as food, clothing and shelter are considered as basic necessities for every citizen, so is electrical energy. The first application of electrical energy was for street lighting which then spread to homes, offices and practically everywhere people lived and worked. Electricity also provided a clean alternative to steam engines in factories for providing motive power. It is a versatile form of energy that can be used for cooking, heating and cooling, appliances and gadgets that are used for cutting down manual labour and improving the quality of service in many spheres of life including health care, education and entertainment. In modern times, electricity drives communication equipment and information technology. The application of electrical drives in rail and road transportation can reduce environmental pollution and improve efficiency of energy usage. About 40 % of the total energy used in developed countries is electrical energy.

We expect the lights to turn on at the flick of a switch and it is quite frustrating when the lights go out suddenly and darkness descends due to unexpected power supply interruptions. In this chapter, we will look into the causes of power blackouts that may be temporary (lasting only few minutes) or prolonged (lasting several hours or even days). The blackout may occur only in a localized area (say in a street, affecting few people) or in a wide area affecting thousands of people. The historic blackout on November 9, 1965 in

New York and New England states in U.S.A affected 30 million people in an area of 83,000 square miles. The consequences of this were to introduce major changes in the grid operation and in energy control centres that saw the application of computers for SCADA (Supervisory Control and Data Acquisition). The computers are also used for security monitoring and to provide guidance to system operators to implement preventive control (to prevent the occurrence of blackouts). However, these measures were applied primarily for bulk power systems made up of generating stations and high voltage transmission networks. Subsequent developments in the eighties, have led to the introduction of Distribution Automation and Control (DAC) for monitoring facilities in the distribution substations (such as transformers and switches) and load management.

1.2 Present Status of Electricity Supply Industry (ESI)

Supplying reliable power at affordable costs has been the guiding principle in planning power (supply) systems, all over the world. Till the nineties, the Electricity Supply Industry (ESI) was a regulated monopoly in which each utility took care of planning, design and operation of the generation, transmission and distribution facilities under its control in the designated area. The tariff was decided on the basis of an equitable rate of return on investment, taking into account the capital and operating & maintenance costs. Such utilities are essentially Vertically Integrated Utilities (VIU). The exchange of energy among VIUs was based on mutual needs with the help of tie lines (interconnections). While most of the utilities in U.S.A. were privately owned, in UK, India and elsewhere, they were publically owned. The desire to introduce competition in generation of power with the expectation of providing a choice for the customers and reducing the costs, led to the unbundling of generation, transmission and distribution. Also, deregulation was introduced where a large customer can buy power from any competing supplier who had access to transmission facilities irrespective of its ownership. The jury is still out on the impact of these momentous changes in the operation of electricity supply industry (ESI). While it was anticipated that the electricity rates will drop without affecting the system reliability, the actual experience showed that it led to higher rates and rolling or rotating blackouts. However, on the positive side, it also opened the door to greener forms of electricity generation (See the article in

December 2010 issue of IEEE Spectrum-"How Free Market rocked the Grid" by Seth Blumsack). In India, the 2003 electricity act has led to the unbundling of generation, transmission and distribution and introduction of open access to transmission facilities and power trading.

In another article (by S.M. Amin published in the January 2011 issue of IEEE Spectrum), entitled "U.S. Electrical Grid Gets Less Reliable", it is stated that the U.S. electrical grid has been plagued by ever more and ever worse blackouts over the past 15 years. "In an average year, outages total 92 minutes per year in the Midwest and 214 minutes in the Northeast. Japan, by contrast, averages only 4 minutes of interrupted service each year." The author states that "Starting in 1995, the amortization and depreciation rate has exceeded utility construction expenditures. In other words, for the past 15 years, utilities have harvested more than they have planted. The result is an increasingly stressed grid." It is also interesting to note that R&D spending for the electric power sector has dropped 74% from a high in 1993 of U.S.$ 743 million to $193 million in 2000. The R &D declined from 0.3% of the revenue in 2000 to 0.17% in 2006.

"Even the hotel industry spends 0.7%. on R&D." Incidentally, a major blackout on August 14, 2003 affected large portions of U.S. Midwest and Northeast, and Ontario in Canada. The power outage affected 50 million people and 61,800 Mega Watts (MW) of load. The power was not restored for 4 days in some parts of U.S. In parts of Ontario, there were rolling blackouts that lasted for more than a week. The estimates of the total costs of the blackout range from $ 6 billion to about $12 billion.

In India, there are no accurate records of the power outages kept. The public seem to consider interrupted power supply as their fate! (Although, it must be admitted that the reliability of power supply in Mumbai had been comparable to that in Europe). It is important to outline the causes that lead to a local or widespread blackout. The basic configuration of an electric power system is shown in Fig.1.1. The electricity is produced in generating stations at voltages from 10 to 25 kilo Volts (kV) from various energy sources such as coal, oil, gas, nuclear, hydo, solar, wind, geothermal and tidal. The electrical power is generated in a three-phase, alternating current (AC) form. The renewable energy sources such as solar and wind are desirable although unsteady and costly. For long distance transmission of power, the generated power is transformed to a higher voltage using a step-up transformer. The transmission voltages range from 220 kV to 800 kV. A 400 kV, three phase AC transmission line can transmit power levels up to 1000 MW. In general, the power rating of an AC line is inversely proportional to its length for reasons that will be mentioned in a later chapter. The maximum rating is determined by the temperature rise of the aluminum conductor that is a function of the ambient temperature and the current carried by the conductor. At higher temperatures the conductor sags and the clearances from the vegetation below the line may become insufficient leading to faults that are cleared by tripping the line. It is possible to transmit power by converting it to direct current (DC) using power electronic converters at the two ends of the line. (It is to be noted that the low voltage (230/400 V) distribution of power is invariably in the AC form).

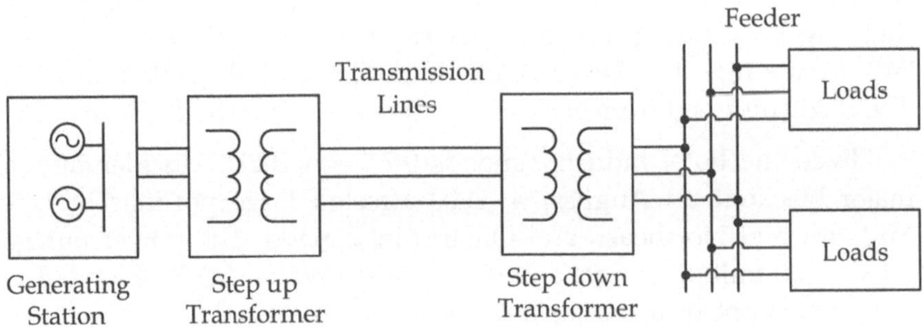

Fig. 1.1 Basic configuration of an electric power system

From cost considerations, the transmission lines are primarily overhead lines that are exposed to vagaries of the nature and susceptible to failures caused by storms and lightning surges. The failures can be of mechanical nature (the worst could be damage to towers) or electrical breakdowns caused by failure of air insulation. Fortunately, the air can recover its insulation strength, once the factors that lead to ionization are absent. Thus, tripping of a line following flashovers permits the air insulation to regain its strength and the line can be reclosed after few seconds. In any case, the grid operation requires the power transfers to be unaffected after the loss of any single line. Thus, the transmission network is a complex electrical circuit containing several meshes (loops).

The distribution network is usually radial with primary feeders operating at voltages of 11 kV to 66 kV (in India) depending on the load (power demand) requirement. Large customers can receive power at 66 kV or sometimes even at 110 kV. The low voltage, secondary distribution network operating at 230/400 V is used to supply residential and small commercial loads. A distribution substation supplies power to feeders by stepping down the voltage by using step down transformers. Further reduction in voltage is provided by pole mounted transformers that supply power to a group of houses or units that may consume power of 3 to 6 kW (peak load) each.

While we will discuss the organization of electricity supply and its operation separately in a following chapter, it is worth noting that electric power utilities were started more than a century ago to primarily supply a local area or a town/city. Historically, the credit for developing electric lights (incandescent lamps) goes to Thomas Edison who also started a power station using DC generators driven by steam engines to supply street lighting in Pearl street in New York city, U.S.A.in 1882. It was soon found that long distance transmission was not feasible unless voltages can be raised. This led to the development of AC generation and transmission as transformers can be built to operate at very high efficiencies and thus, stepping up or down of voltages became feasible. It is to be noted that for a given power level, the current flow in a line is inversely proportional to the voltage level. The power losses in a conductor are proportional to the square of the current and hence the combined losses in the transmission and distribution networks can be kept below 10% of the power generated. AC generators could be built also for high power ratings unlike DC generators where commutator and brushes limit the power level.

The discovery of AC induction motor by Nicola Tesla also spurred the utilization of power in the AC form. As the power utilities increased in number, it was found desirable to interconnect them using tie lines for improved economy and reliable operation. (It is to be noted that there is no difference physically between a line evacuating power from a generating station and a tie line. In the latter, the power can flow in both directions). This has led to the growth of large interconnected systems, spread over a large geographic area. In India, there are five regional grids-Northern,

Southern, Western, Eastern and North Eastern. Except for the Southern regional grid, all other regional grids have now been connected using AC tie lines. Even the Southern regional grid has been connected to the Western and Eastern grids using DC ties that don't have the problems associated with the AC ties. However, DC ties are more expensive than AC ties in general. In North America, only Texas State in U.S.A. is not interconnected (except using DC ties). The rest of the states/provinces are organized into two major Interconnections-Western and Eastern. These two major areas are also connected by small DC ties.

1.3 Certain Facts that Govern Electricity Supply

Before we analyze the causes that can result in blackouts (localized or widespread) and the remedies, let us look at some important facts that affect the planning, operation and control of interconnected AC power systems.

FACT 1: At any instant of time, there is a power balance-that is, the power supplied to customers = power generated – losses in the system.

FACT 2: It is not economically feasible to store electric energy in large quantities.

FACT 3: It is not possible to distinguish between the energy received from different generating stations. (The concept of green (renewable) power is only conceptual in the sense that nobody can claim that they receives only green power unless they are not connected to the grid).

FACT 4: The power flow in an AC line cannot be controlled unless special control equipment is used.

FACT 5: The electricity travels over transmission lines at speeds approaching that of light (300,000 kilometers/second).

FACT 6: The rate at which power can be ramped up is limited by prime-mover characteristics.

Let us first discuss the implications of these facts. While economists wish to consider electrical energy as a product and encourage competition among generators, and power trading, FACT 3 indicates that the products of all manufacturers (generators, in this case) are identical. Also, power delivered to a distribution substation generally comes from more than one generating station and their contribution is determined from the network configuration and Kirchhoff's laws (that determine flows of current in individual lines).

FACT 1 and FACT 2 indicate that variations in the load (power demand) require prompt adjustments in the generated power to maintain power balance without affecting the frequency. FACT 5 implies that the adjustments in the generation have to be done instantaneously to avoid fluctuations in the frequency. However, FACT 6 indicates that this is not feasible. It is to be noted that a sudden increase in load results in deceleration of the rotors of generators and a drop in the frequency. We can also say that the increased energy demand is initially met by the stored kinetic energy in the spinning generator rotors. The speed governor action results in opening of the valves (carrying steam or water to the steam or hydro turbines) causing increase in the generated power. To restore frequency to the original (nominal) value, the power setting has to be changed to the desired new value determined from Automatic Generation Control (AGC), also called as Load Frequency Control (LFC). A large increase in the load requires adequate spinning reserve at the generators already in service.

The system consisting of generating stations and high voltage transmission network is generally called as bulk power system. The load on the system is determined by the customer requirements and has typically, a cyclic variation during a period of 24 hours. The load is generally minimum during the nights and picks up during the day reaching a maximum value which can be more than twice the minimum value. There could also be more than one peak (one in the morning and other in the evening). The maximum load can also vary depending on the seasonal variations, holiday periods etc. However, the trend is for the peak daily load to increase with time as the economy and population expands. The annual load growth is generally high in a developing country like India (7-10%). If energy storage in large quantities is feasible, it is adequate to plan generation facilities to provide only average load on the system rather than the peak load. The blackouts in India are primarily due to the FACT 2 and the installed capacity is perennially insufficient to meet the peak load.

Apart from the normal variations in the load (that are represented by a daily load curve), there can be disturbances caused by sudden tripping of generators and transmission lines, connection/disconnection of large loads. This will affect normal power flows in the lines and it is possible that some lines can be overloaded. (It is to be noted that the power flow in transmission lines is determined by injections of power at generating and load nodes in addition to Kirchhoff's laws). To prevent abnormal rise in the conductor temperature, it is necessary to trip the line. Since all lines are not overloaded (some lines operating with sufficient margins), it would be desirable to control power flows in the overloaded lines to ensure they don't trip. However, FACT 4 shows that this is not possible unless special control equipment is provided which involves additional investment (although it may be cheaper than putting up a new line). Interestingly such controllers (based on power electronics) are called as FACTS Controllers. Here, FACTS is an acronym for Flexible AC Transmission System.

It is to noted that from the customer's point of view, the major facet of Power Quality (that is assuming importance) is reliability (availability of uninterrupted power). In addition, regulation of voltage and frequency is also significant. The speed of regulation of voltage (how fast the voltage is brought back to the normal value) can be important for critical loads such as paper mills and semiconductor manufacturing plants.

Additionally, the deviation of the voltage waveform from the (ideal) sinusoidal can affect motor loads.

1.4 Causes for Lights Going Out

1. ***Inadequate Spinning Reserve***: The spinning reserve takes care of the contingencies involving sudden tripping of a generator or connection of a large load. Normally, load variation tends to be slow as the total load is made up of a large number of small loads and there is diversity among them. Typically, the size of the spinning reserve should not be less than the rating of the largest generator in service. If the spinning reserve is inadequate, the loss of a large generator can result in decay of frequency which is arrested by frequency activated load shedding (to restore the balance between the supply and demand).

   In developing countries, the installed generation capacity may not be adequate to meet the peak load. In such cases, it is necessary to resort to rolling (rotating) blackouts where the supply to a set of customers is deliberately switched off for the required duration to reduce the peak load. Even if the installed generation capacity is adequate, there could be shortage (non-availability) of fuel (fossil or nuclear) that results in power cuts. Also, in hydroelectric power plants, there could be deficiency of storage in reservoirs due to draughts or insufficient rain fall in catchment areas.

It is to be noted that even under generation shortage conditions, it is important to maintain the required minimum spinning reserve. While it may be tempting for the system operator to have zero reserve to avoid customer complaints, this is not advisable as most of the customers would plan for scheduled outages, rather than be rudely surprised by the sudden random outages that can lead to power blackouts.

2. ***Inadequate transmission capacity***: For the specified locations of generators and loads, there is an upper limit on the aggregate load that can be supplied by a transmission network. For a specified loading condition and the power flow in a transmission line, it is possible to estimate the Available Transfer Capacity (ATC) that cannot be exceeded. This implies that even if the generation is adequate to meet the required load, the transmission is inadequate. In the economist's language, it is called 'congestion' (as the transmission network is compared with the transportation network). However, while the transportation network is subjected to only KCL (Kirchchoff's Current Law), the electrical transmission network is also subjected to KVL (Kirchchoff's Voltage Law) which puts additional constrains. For example, when an overlay of EHV lines is designed to carry the required power (for a 765 kV line of 150 miles length, 3800 MW can be transmitted if there are no other constraints), it has been observed that underlying lower voltage lines carry higher power because of the increase in the impedance of the EHV line due to the autotransformers at the two ends of the line used for raising the voltage (say, from 345 kV to 765 kV).

3. ***Faults (Short Circuits) in the Distribution Feeders***: The breakdown of insulation in cables or overhead distribution feeders (at medium voltage levels) result in interruption of power flows to loads. Since the faults have to be cleared by opening of circuit breakers, the supply is affected until the circuit breaker is reclosed after the fault is cleared. For overhead conductors, the insulation is provided by

the air which can break down due to lightning or insufficient clearance to a grounded conductor (of electricity). However, when the causes resulting in a fault are removed, the air insulation recovers and hence the breaker is reclosed after some time (of clearing the fault). If the fault is still persisting, 3 or more attempts are made to clear the fault and reclose after some time delay. For a permanent fault, the restoration of the supply requires another healthy feeder (to which the load can be connected).

4. ***Major Disturbances in the System***: A power system is normally designed to withstand a single major contingency. However, unforeseen events can occur that can result in cascading outages (tripping of several lines and generators due to overloads and under frequency or loss of synchronism) that causes major blackouts which can be widespread over a large geographical area. (as one that happened on 14th August 2003 in Canada and USA or the one that happened on 31st July 2012 in India).

5. ***Improper Coordination of Protection***: An electrical circuit, even at low voltages (110 or 220 V) and carrying 10 amperes has to be protected against faults (short circuits resulting from failure of insulation). In the past, fuses that would blow at high currents resulting from faults, were used. Now a days, miniature circuit breakers (MCB) and earth leakage circuit breakers (ELCB) are used for protection. In the absence of such protection, the fault currents would continue to flow in the distribution wires and this can result in the pole fuses blowing out, resulting in the interruption of power to several customers. In the worst case, the breaker/recloser at medium voltages (11 to 33 kV) can operate inconveniencing a large number of customers in a locality.

   It should be noted that the protective device at a customer premise must be carefully chosen based on the connected load and the peak load demand.

6. ***Fluctuating Power Generation with Renewable Energy Sources (RES)***: Both wind and solar power are useful and viable renewable energy sources. However, wind velocity fluctuates and the power output of a wind turbine is proportional to the cube of wind velocity. Thus, a 10 % reduction of rated wind speed results in 30 % reduction of power output. Also, there is a cut-out wind speed (20 to 25 m/sec) above which the wind turbine needs to shut down

to protect the turbine. Similarly, photo voltaic (PV) solar power generation suffers from the variation of sunshine due to clouds and earth's rotation. The only solution to the variable RES is to install adequate electric energy storage (EES) facilities. However, this adds to the cost of generation. At present, several types of EES technologies are under development.

1.5 Reliability of Electricity Supply

For the customers connected to low or medium voltage feeders, the reliability indices are the frequency and duration of the power outage. We can also define average service availability index (ASAI) which considers the total duration of power interruptions in a year. In U.S.A. prior to restructuring, the best index was 1 hour in a year (which translated to 99.99% availability). Of course, in the present times when lighting is only one of the various uses of electrical energy, the power quality (PQ) which encompasses issues such as frequency, slow or fast variations of voltage magnitude (sags and swells), voltage flicker ad departure from sinusoidal waveform (that results in the harmonics which can affect motor loads) is becoming important. In the past, steady state voltage magnitude remaining within a band (+/- 5 % of the nominal) and frequency close to the nominal were considered adequate.

Power Quality (PQ) is fast becoming a major issue in power distribution. One of the PQ problems is the generation of harmonic currents by nonlinear loads caused by introduction of power electronic control for efficient energy utilization. The presence of harmonics (resulting in the departure of the current wave form from being sinusoidal) causes increased losses in the lines and motors. The presence of harmonics injected by a load results in the distortion of the voltage wave form at other locations due to the impedance of the distribution lines. Active filters based on reactive power compensation principle are being applied to overcome this problem. The fast variations in the voltage magnitude (including momentary interruptions of power) are also deleterious to the critical and sensitive loads such as paper and steel mills and semiconductor fabrication plants.

1.6 Scope of this Book

The basics of DC and AC circuits (both single phase and three phase) are introduced in chapter 2 with definitions of three circuit elements-resistance, inductance and capacitance. This chapter is intended for beginners and can be skipped by those familiar or not interested in circuit analysis. The electrical quantities are defined mathematically in this chapter.

The basic structure of the electricity supply system (also called as electrical power system in technical parlance) including generation and delivery of power, and the normal control functions during system operation are explained in chapters 3 to 5. The protection against faults (short circuits) mainly caused by lightning discharges affecting overhead lines, is described in chapter 6 along with the collapse of the system caused by contingencies that lead to the system transition to an emergency state where the system operator has no time to react in real time. Chapter 7 describes the various technological solutions to enhance system security and eliminate widespread blackouts. The emphasis is to enable the creation of self-healing transmission grids.

In this context, the present developments on smart grids which are mainly aimed at low or medium voltage microgrids is presented with a discussion on their relevance in improving the system operation that should provide desired reliability without sacrificing cost effectiveness.

Chapter 8 presents various types of renewable energy sources (RES), the predominant being wind and solar. Since their application also involves the use of energy storage technologies, these are discussed in this chapter. Finally, the issues of restructuring of electrical supply industry (ESI) and competition are discussed in detail with some suggestions for system planning in the present regime of reregulation or restructuring.

The Appendices cover the unit system in general, the electrical shock and safety issues, nuclear energy and glossary.

CHAPTER 2

Basics of Electrical Circuits

"By a model is meant a mathematical construct which, with the addition of certain verbal interpretations, describes observed phenomena."

- John Von Neumann

2.1 Introduction

In this chapter, we briefly review the basics of electrical circuit theory that is relevant in understanding the working of electricity supply. The terms of charge, voltage, current, power, reactive power and power factor will be explained. The basic circuit elements are resistance, inductance and capacitance. The reader who has studied high school physics may recall that most of the material covered here is included in that course.

The electric charge is a very basic concept in the study of electric fields. There are two types of charge - one is called positive and other negative. The electrons in an atom have negative charge and atoms/molecules in which electrons have been dislodged (due to application of electric field) carry positive charge. A voltage source such as a battery creates an electric field in which a charge moves due to the force applied. The force (F) is given by

$$F = qE \qquad .....(2.1)$$

where q is the magnitude of the charge expressed in coloumbs in SI units(see Appendix A) and E is the electric field measured in volts/meter. Assuming the electric field is uniform over the space in which the charge moves, the work done in traversing a unit distance is the product of voltage and charge.

The electrical circuit theory is concerned with the analysis of current flows in electric networks made up of components. There are three basic passive elements - resistor (R), inductor (L) and capacitor (C). Series and parallel connections of these elements form a passive network which can be excited by applying voltage sources, which result in flow of currents in the individual element/component of a network. Complex equipment/component such as generators, transformers, transmission lines or motor loads can be modeled by equivalent circuits with passive and active elements. The passive elements do not generate or supply energy. They can only dissipate or store energy supplied from active elements. Voltage and current sources are the active elements.

2.2 Circuit Elements and Sources

A voltage source, in general, produces a time varying voltage across the two terminals to which it is connected. An ideal voltage source is one where the voltage is independent of the current supplied. In practice, this does not apply and the voltage tends to drop (see Fig. 2.1(a)) as the current supplied is increased. Note that we are assuming here a DC voltage source that produces a DC current. The characteristics shown in Fig 2.1(a) can be modeled by an ideal voltage source in series with a fictitious resistor (R_i) whose value is negative slope of the characteristic.

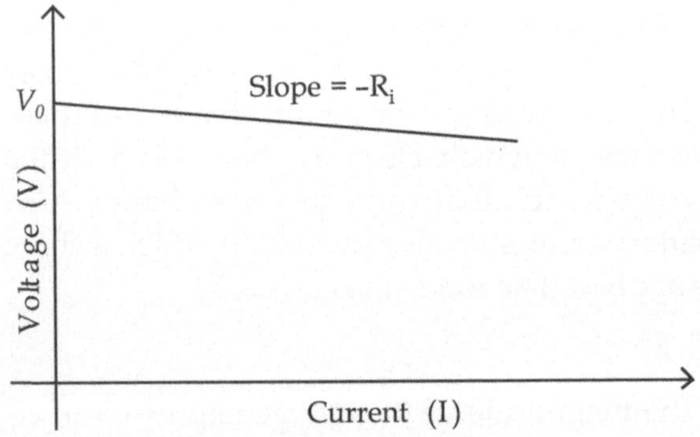

(a) V – I Characteristic

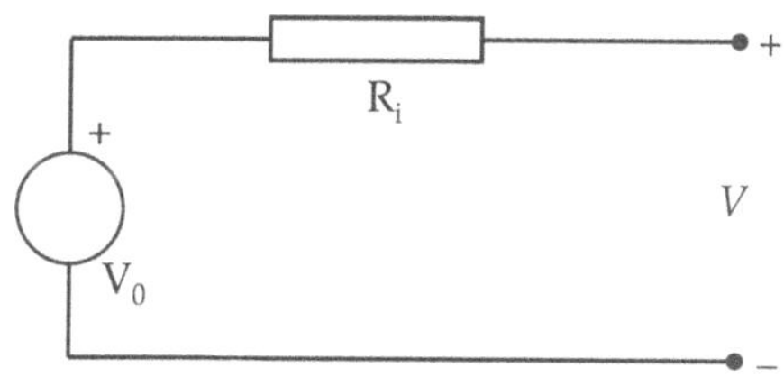

(b) Equivalent circuit

Fig. 2.1 A DC voltage source

It is also possible to define an ideal current source which maintains the current independent of the voltage across it. The circuit symbols for ideal current source and the passive circuit elements (R, L and C) are shown in Fig. 2.2. The symbols for the passive elements is not always universal and when there is no doubt, a simple symbol (rectangle) can also be used for R and L.

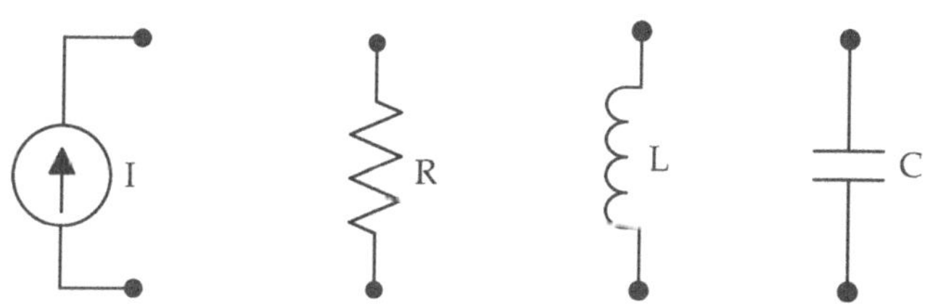

(a) Current source **(b)** Resistor **(c)** Inductor **(d)** Capacitor

Fig. 2.2 Circuit symbols

The resistor (R) dissipates energy (in the form of heat). The ideal inductor (L) can only store energy in the magnetic field created by the flow of current through it. The ideal capacitor (C) also stores energy in the electric field created by the application of voltage across it. Resistance, inductance and capacitance are the measures of the resistor, inductor and capacitor respectively. Since same symbol (R) is used to denote a resistor as a network element and also its resistance value, it is convenient to term R as resistance. Similar comments apply to the symbols L and C.

If we apply a DC voltage source across a resistance (R), then the current is given by

$$I = \frac{V}{R} \qquad \text{.....(2.2)}$$

The current (I) is established as soon as the voltage is applied and remains constant. The power loss (P) is given by

$$P = I^2 R = \frac{V^2}{R} \qquad \text{..... (2.3)}$$

The units for voltage, current and power are Volt (V), Ampere (A) and Watt (W) respectively. The unit for resistance (R) is in Ohm (note all these units are named after well known scientists).

The incandescent electric light and heater coil are examples of resistance. However, it is to be noted that their resistance (R) is a function of the temperature. It increases with the temperature and thus, the current through a light bulb is maximum at the instant of switching on. It is observed that a light bulb fails typically when it is switched on rather than when it is burning.

2.3 Energy Storage Elements

Inductor

Inductance (L) and capacitance (C) are both energy storage elements. An inductance creates a magnetic field when a current flows through it. The energy (W) is stored in the magnetic field given by

$$W = \frac{1}{2} L I^2 \qquad \text{.....(2.4)}$$

where W is the energy in joule (watt-second), L is inductance in henry and I is current in ampere. Consider a DC voltage source connected across an inductance through a switch (S) as shown in Fig. 2.3 (a)

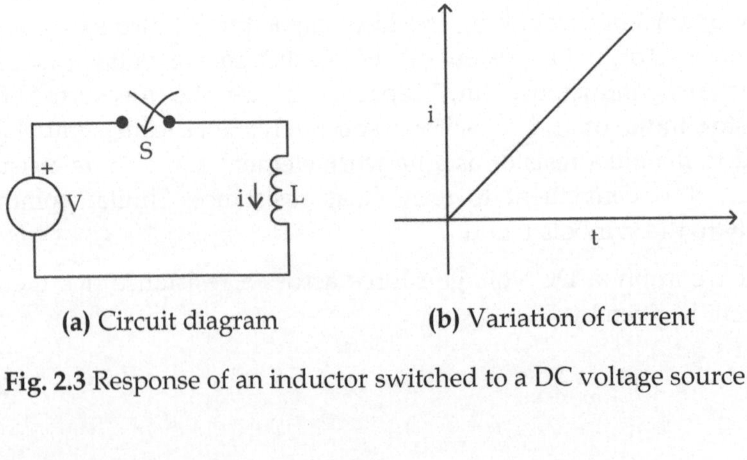

(a) Circuit diagram **(b)** Variation of current

Fig. 2.3 Response of an inductor switched to a DC voltage source

If the switch is turned on at t = 0, the current (*i*) in the inductance (*L*) increases linearly as a function of time (*t*) (see Fig.2.3 (b)). Theoretically, the current can increase without any limit, in practice there is some resistance (say R) associated in an inductive circuit and the maximum current will be given by V/R as before.

The main thing to note is that current in an inductive circuit cannot jump to the final value instantly. It has start from the initial value of zero. The final value is reached after some time depending on the time constant, $T = \frac{L}{R}$.

Note: Mathematically, the final value is reached after a very long time (theoretically infinite!) however, 95% of the final value is reached after t = 3T.

Since the final value of current is independent of L, we can state that L can be ignored (short circuited) to compute the final value in a DC circuit. However, this is not valid in AC circuits which will be considered later.

Any conductor carrying current produces magnetic field and thus has inductance. Generators and motors require magnetic field to generate voltage or produce torque on the rotor. Ignition coil in an automobile is also a typical example of an inductor. Breaking current in an inductive circuit provides a spark as a result of high voltage across the gap caused by the break in the circuit.

Capacitor

A capacitor stores electric charges and consequently, energy. The energy stored in capacitor (C) is given by

$$W = \frac{1}{2} C V^2 \qquad \text{.....(2.5)}$$

where W is the energy in joule, C is the capacitance in farad and V is the applied voltage in volt.

Consider a capacitance (C) connected to a DC voltage source (V) though a switch S (see Fig. 2.4 (a)). The switch is initially off and the capacitor has zero initial charge when the switch is turned on at t = 0.

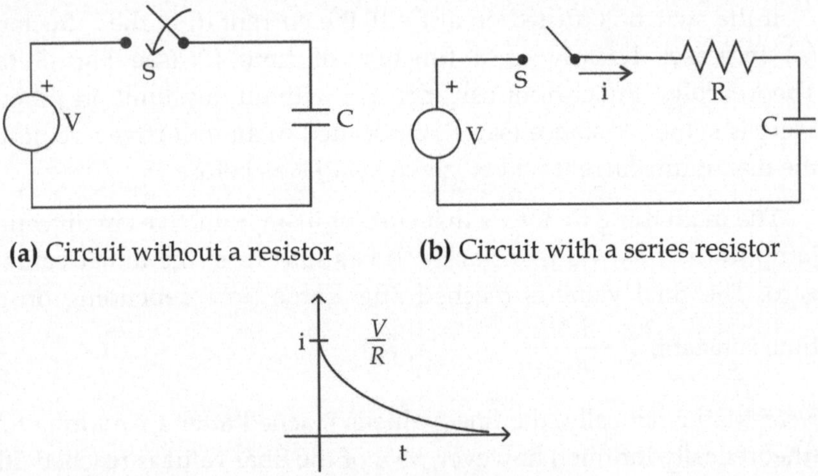

(a) Circuit without a resistor **(b)** Circuit with a series resistor

(c) Variation of current

Fig. 2.4 Response of a capacitor switched to a DC voltage source

A surge of very high current flows through the capacitor to charge the capacitor practically instantaneously to attain the final value of V. The total charge 'q' when the capacitor is fully charged is given by

$$q = CV \qquad \text{.....(2.6)}$$

The final value of the current is zero. To limit the initial current, it is necessary to connect a resistance (R) in series with the capacitor. The initial current is now V/R and it decays exponentially with time. The final value is zero when the capacitor is fully charged.

Based on the observations regarding energy storage in an inductor or capacitor, it can be said that a resistor of suitable value has to be connected to limit the peak current flowing through L or C. The only difference is that peak current is of transient nature in C while it is steady in L. This implies that choice of R is critical in charging an inductor.

Capacitors are used for power factor correction or improving the voltage profile in AC circuits. Transmission lines also have capacitances between conductors and between the conductor and the ground.

The stored energy in L or C can only be dissipated in a resistor and cannot be returned to the DC voltage source unless complex power electronic converters are utilized. Trying to switch off current in an inductive circuit can be result in high voltages across the switch contacts as they separate. This in turn, can result in an arc due to break down of the air. In contrast, switching off a capacitive circuit does not cause any problems. However, short circuiting a charged capacitor can result in dangerously high currents that can damage the capacitor.

2.4 AC Circuits

Generally, we consider electric circuits as networks of passive elements, excited by voltage and current sources. When these sources are constant (do not vary with time), we say the circuits are DC. If the sources are varying with time (sinusoidally) we say the circuits are AC. AC circuits can be single phase or three phase. While generation and transmission are invariably three phase, distribution at low voltage is generally single phase with two wires (or conductors). The return conductor is the neutral wire that is normally maintained at ground potential.

Single Phase AC Circuits

Consider a single phase AC circuit shown in Fig. 2.5

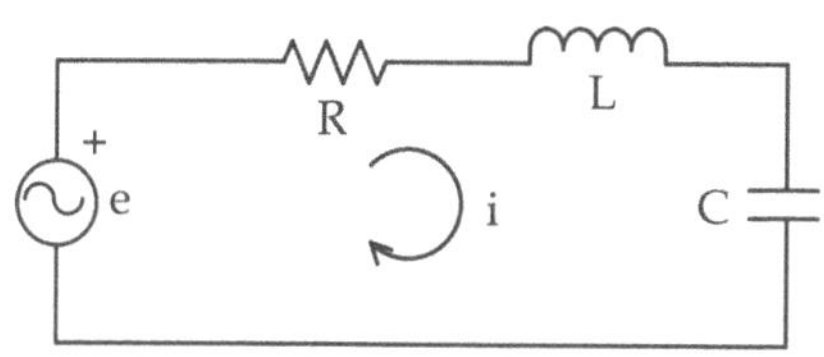

Fig. 2.5 Single phase AC circuit

The voltage source e is a function of time given by

$$e(t) = E_m \sin \theta, \ \theta = \omega t + \alpha \qquad .....(2.7)$$

where E_m is the maximum or peak value (also called amplitude), ω is the radian frequency, α is a constant angle (termed as phase angle). Note that any sinusoidally varying quantity is defined by three quantities - (a) magnitude (b) frequency and (c) phase angle with respect to a standard sine wave that phases through zero at time t = 0. The variation of voltage with time is shown in Fig. 2.6.

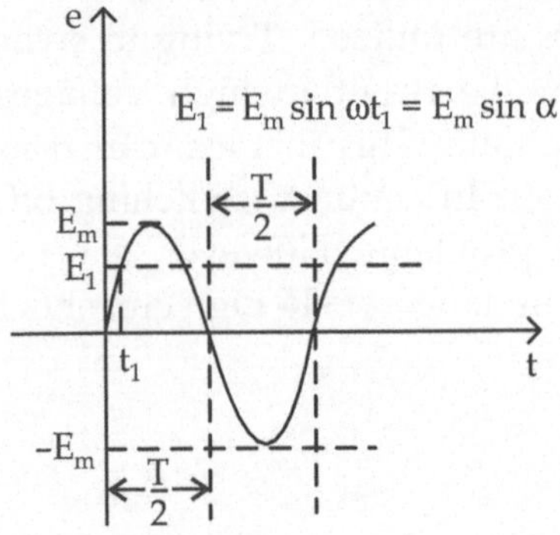

Fig. 2.6 Variation of Voltage with time

It can be observed that the function is periodic (it repeats after a period of T). the sine function is also symmetric about the x (time) axis; that is

$$e\left(t+\frac{T}{2}\right) = -e(t) \quad .....(2.8)$$

It can also be shown that

$$e^2(t) + e^2\left(t+\frac{T}{4}\right) = E_m^2 \quad .....(2.9)$$

This follows from the fact that

$$sin^2\theta + cos^2\theta = 1 \quad .....(2.10)$$

and

$$e\left(t+\frac{T}{4}\right) = E_m cos\theta \quad .....(2.11)$$

It can be shown that the period $T = \frac{1}{f}$ where f is the frequency expressed in cycles per second. Actually the unit for frequency is named after the physicist Hertz and is denoted as Hz. The relationship between ω (radian frequency) and f is given by

$$\omega = 2\pi f \quad .....(2.12)$$

where π is the ratio of the circumference to the diameter of a circle

It is to be noted that sine and cosine of an angle (in a right angled triangle) are defined in high school trigonometry. Consider a right angled triangle shown in Fig. 2.7 with one side (OA) of the triangle drawn along the x-axis.

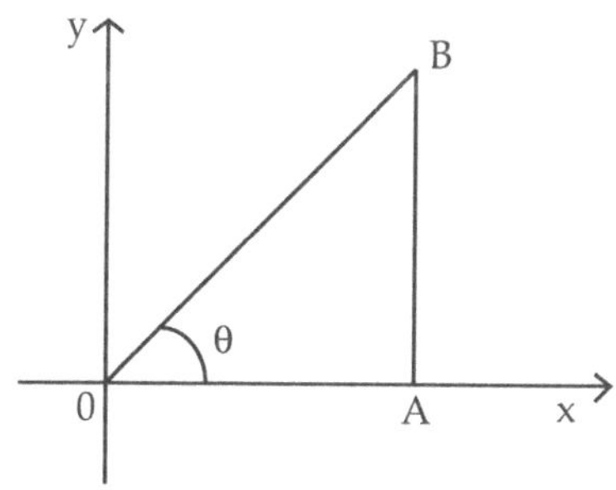

Fig. 2.7 A right angled triangle

θ is the angle between OB (the hypotenuse) and OA. Sine and cosine of θ are defined as

$$sin\theta = \frac{AB}{OB}, \ cos\theta = \frac{OA}{OB}$$

From Pythagoras theorem stating that $OA^2 + AB^2 = OB^2$, we get the result given by equation (2.10).

If θ is varying linearly with time, we can observe the variation of sin θ by assuming the line OB (termed as a vector) rotating about the origin O, with uniform speed of ω radians per second. The vector will complete one revolution in T seconds. When $\theta = 0$, $AB = 0$ and hence sin $\theta = 0$, when $\theta = \frac{\pi}{2}$, sin $\theta = 1$. Similarly, it can be observed that $sin\pi = 0$ and $sin\frac{3\pi}{2} = -1$.

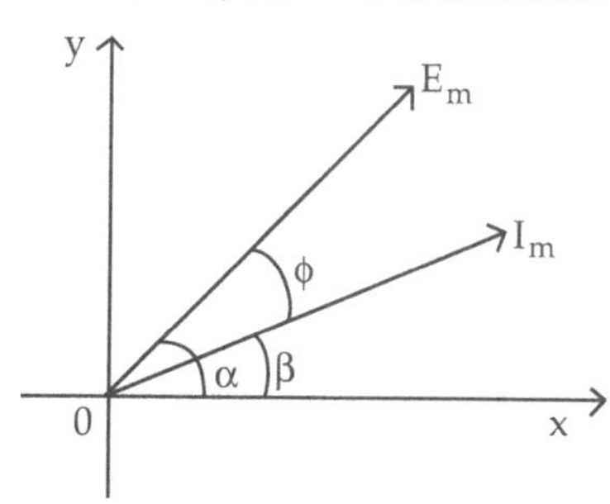

Fig. 2.8 Phasor diagram

It is interesting to note that a sinusoidally varying function can be represented by a rotating vector and its projection on the y-axis gives the instantaneous value. If there are two sinusoidal functions with the same frequency (*f*), they can be represented by two vectors rotating at the same speed. For example, the voltage *e* and current *i* in Fig. 2.5 are shown by the two vectors denoted by E_m and I_m (see Fig. 2.8).

The position of the vectors is shown at $t = 0$. This indicates the current i in the circuit is given by

$$i(t) = I_m sin(\omega t + \beta) \quad .....(2.13)$$

Note that, what matters is the magnitudes of the two vectors and the relative phase angle (ϕ) of the two vectors. This information is preserved even if consider the position of the vectors at some other time.

By convention, the magnitude of a sinusoidal quantity (voltage or current) is not represented by its peak value, but its Root Mean Square (RMS) value. Mathematically, this is defined as

$$F = \left[\frac{1}{T}\int_0^T f^2 dt\right]^{\frac{1}{2}} \quad .....(2.14)$$

where f can represent e or i. It can be shown that

$$E = \frac{E_m}{\sqrt{2}}, I = \frac{I_m}{\sqrt{2}} \quad .....(2.15)$$

where E and I are the rms values of voltage and current respectively.

Note:

1. The significance of RMS value of voltage is that if a sinusoidal voltage of magnitude E is connected across a resistance one ohm; the power consumed is E^2. Similarly, if a current of magnitude I is flowing through a resistance of one ohm, the power consumed is I^2.

2. The definition of RMS value given by equation (2.14) is unaltered even if the function is not sinusoidal. The only restriction is that it should be periodic function of period T.

2.5 Mathematical Representation of the Rotating Vector

Referring to Fig. 2.8, we note that the two vectors for the voltage and current are rotating counter-clockwise at constant speed of ω radians per second (1 radian = 57.3^0) or 360 f degrees per second. We can express the instantaneous values of the voltage and current as

$$e(t) = E_m sin(\omega t + \alpha) = Im\left[E_m e^{j\theta}\right] = Im\left[\left(E_m e^{j\alpha}\right)e^{j\omega t}\right],$$

since $\theta = \omega t + \alpha$

$$i(t) = I_m sin(\omega t + \beta) = Im\left[\left(I_m e^{j\beta}\right)e^{j\omega t}\right]$$

$e^{j\theta}$, $e^{j\alpha}$, $e^{j\beta}$ and e^{jwt} are complex numbers. α and β are constant phase angles; Im indicates imaginary part of the complex number. $e^{j\theta}$ is defined as

$$e^{j\theta} = cos\theta + jsin\theta$$

Note also that $j = \sqrt{-1}$ is an imaginary number.

We define the voltage vector $(\overline{E})$ and the current vector $(\overline{I})$ as

$$\overline{E} = Ee^{j\alpha} = E \cos \alpha + jE \sin \alpha \quad .....(2.16)$$

$$\overline{I} = Ie^{j\beta} = I \cos \beta + jI \sin \beta \quad .....(2.17)$$

Here, $E = \frac{E_m}{\sqrt{2}}$, $I = \frac{I_m}{\sqrt{2}}$ are RMS values of the voltage and current respectively (as explained earlier).

Note:

1. The voltage and current vectors are unlike the vectors used in mechanics, where force and torque are also defined as vectors. To distinguish the AC quantities from vectors defined in mechanics, they (voltage and current) are referred as phasors represented by complex numbers.
2. The phasors shown in Fig. 2.8 are snap shots of the rotating vectors taken at $t = 0$. However, there is no significance to the choice of the instant when the snap shot is taken. Hence, we can assume $\alpha = 0$. In this case, $\beta = -\phi$.

Relationship between Voltage and Current Phasors

For the circuit shown in Fig. 2.5, we can choose $\overline{E}$ as the reference phasor with α=0 or $\overline{E} = E + j0$.

The current phasor $\overline{I}$ is related to $\overline{E}$ by

$$\overline{I} = \frac{\overline{E}}{\hat{Z}} = \frac{E\angle 0}{|\hat{Z}|\angle\phi} = \frac{E}{|\hat{Z}|}\angle -\phi$$

where $\hat{Z}$ is the (complex) impedance. This is not a phasor but is represented by a complex number. $|\hat{Z}|$ is the magnitude of the impedance and ϕ is the angle of the impedance (expressed in polar form). In rectangular form $\hat{Z}$ is expressed as

$$\hat{Z} = R + jX, X = j\left(\omega L - \frac{1}{\omega C}\right)$$

where X is the reactance of the circuit in series with the resistance R.

The magnitude (Z) of the complex impedance and angle ϕ are given by

$$Z = \left(R^2 + X^2\right)^{\frac{1}{2}}, tan\,\phi = \frac{X}{R}.$$

Fig. 2.9 shows the impedance triangle giving the relationships between R, X and Z.

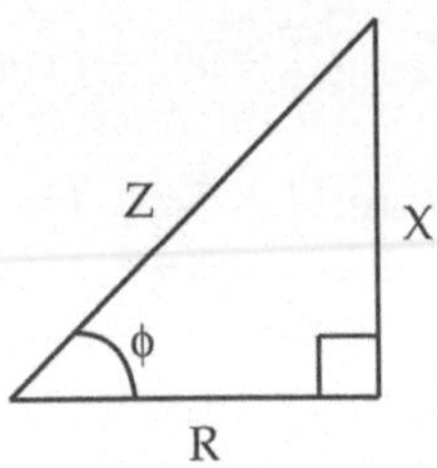

Fig. 2.9 Impedance triangle

Note that the real part of the complex impedance is R (Resistance) which is defined earlier. The imaginary part of the impedance is called as reactance. The reactance has contribution from both inductor and capacitor. While the inductive reactance is positive and directly proportional to the frequency, the capacitive reactance is negative and is inversely proportional to the frequency. An AC transmission line has series inductance and shunt capacitance.

2.6 Power in AC Circuits

In DC circuits there is only a single definition of power *(P)* which is a product of voltage *(V)* and current *(I)*. However, there are several definitions applicable in AC circuits. These are described below:

Apparent Power *(S)*

The product of voltage magnitude (RMS), V and the current magnitude (RMS), I is termed as apparent power. While the dimension of apparent power is same as that of power, the units are labeled as Volt Amperes (VA). The larger units are Kilo VA (KVA) and Mega VA (MVA) (thousand and million VA respectively).

The apparent power is larger than power as will be shown. It is useful in denoting the rating (size) of AC equipment. For example generators and transformers are rated in tens or hundreds of MVA.

Instantaneous Power, Average Power and Power Factor

If we express the instantaneous values of voltage and current as

$$v(t) = \sqrt{2}\, V \sin wt, \quad i(t) = \sqrt{2}\, I \sin(wt - \phi)$$

The instantaneous power *(p)* is a product of the instantaneous voltage and current. For $V = 1, I = 2, f = 50$ and $\phi = 30^0$, p is shown in Fig.2.10. It is observed that p has an average component and a component that oscillates at twice the system frequency. The average power (P) is the product of (RMS) voltage, (RMS) current and a quantity called as power factor(PF) defined by

$$(PF) = \frac{P}{VI} \qquad \text{.....(2.18)}$$

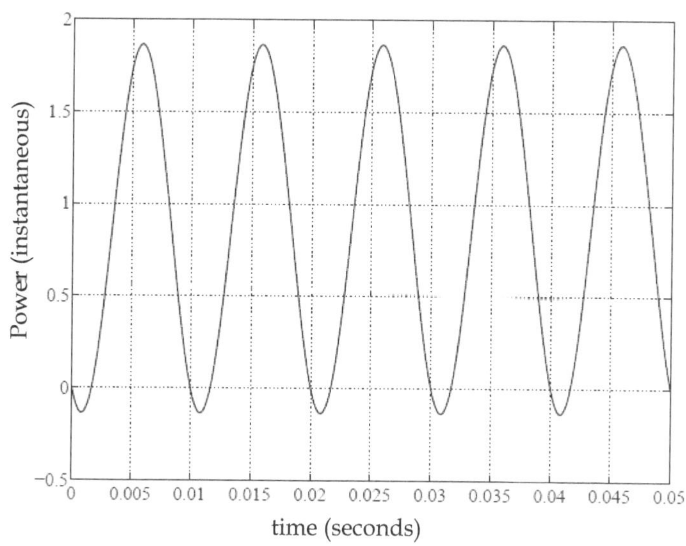

Fig. 2.10 Variation of instantaneous power

For sinusoidal voltage and current, PF is also computed as $\cos\phi$ where ϕ is the phase angle difference between the voltage and current. Since $\cos\phi \leq 1$, we note that $P \leq S$. If the current is lagging the voltage ($\phi > 0$), we say the PF is lagging. If current is leading the voltage, ($\phi < 0$), we say that PF is leading. Generally, the system loads (e.g. motors) have lagging power factor as their reactance is of inductive nature.

To give an example of the power relations, a large AC generator has rating of 555MVA, rated PF of 0.9 and rated power of 500 MW.

Reactive Power (Q)

In Fig. 2.10, it is observed that there are two intervals during a cycle when the instantaneous power is negative. What this implies is the energy is flowing backwards from the load to the source. If PF is unity, the load is purely resistive and the power consumed is always positive (or zero when e and i are zero). The possibility of the negative instantaneous power arises when reactance is present. We know that an ideal inductor or capacitor does not consume, but only stores energy. If we consider a purely reactance type load the average power (P) will be zero and the load will draw power from the voltage source (generator)and return it as the energy flow is reversed. We can conceptually separate the load into two parallel connected parts - one, a pure resistance and the other, a pure reactance (see. Fig. 2.11)

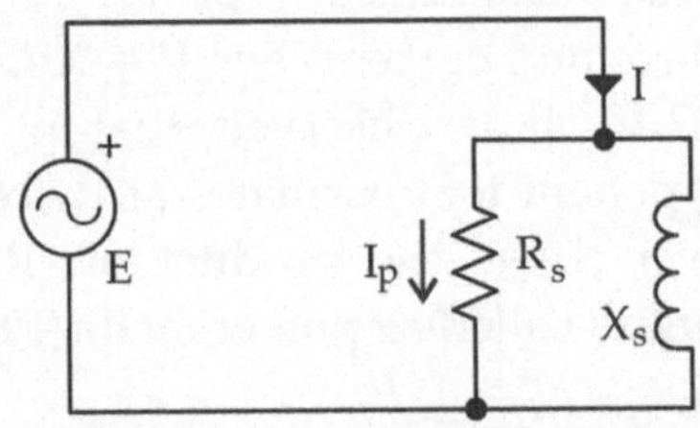

Fig. 2.11 AC circuit with parallel connected resistance and reactance

It can be shown that power $(P) = I_p^2 R_s$ where $R_s = \dfrac{R^2 + X^2}{R}$ and $I_p = I \cos\phi$ is called active current. The quantity $I_q^2 X_s$ is defined as reactive power (Q). $Iq = I \sin\phi$ is called the reactive current. The parallel (or shunt) connected reactance, $X_s = \dfrac{R^2 + X^2}{X}$.

Note that $I_p^2 + I_q^2 = I^2$ and hence we get,

$$S^2 = P^2 + Q^2 \qquad \text{.....(2.19)}$$

The reactive power (Q) is also defined as

$$Q = EI_q = EI \sin\phi \qquad \text{.....(2.20)}$$

Note: The current phasor $\bar{I}$ is expressed in two ways

(i) $\bar{I} = \dfrac{\bar{E}}{\hat{Z}}$ or (ii) $\bar{I} = \hat{Y}\bar{E}$

where $\hat{Y}$ is complex admittance defined as

$$\hat{Y} = \frac{1}{\hat{Z}} = G + jB = \frac{1}{R + jX} \quad .....(2.21)$$

G is called conductance and B is called the susceptance. Note that $G \neq \frac{1}{R}$ and $B \neq \frac{1}{X}$. Also, B is negative if X is positive and vice versa.

2.7 Three Phase AC Circuits

The supply of power at medium and high voltage is invariably by three phase lines. Here, each phase has a sinusoidal voltage source normally connected between the line and a common reference terminal (typically connected to ground) as shown in Fig.2.12. Here, the phases are labeled as

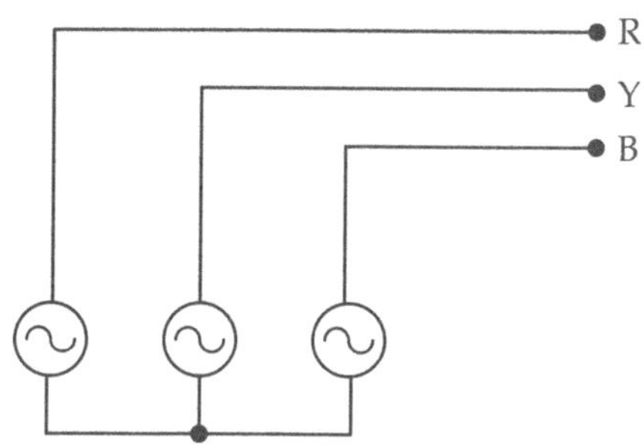

Fig. 2.12 A three phase star connected voltage source

R (Red), Y (Yellow) and B (Blue) following a colour code. Sometimes they are also labeled on A, B and C for convenience. For a balanced supply, the voltage magnitudes in all 3 phases are identical and their phase angles are displaced from each other by $120°$.

We also define a (positive) phase sequence (say RYB) where phase 'Y' voltage lags phase 'R' voltage by $120°$ and phase 'B' voltage lags phase 'Y' voltage by $120°$ The vector or phasor diagram of balanced voltage V_R, V_Y and V_B (with phase sequence RYB) is shown in Fig.2.13(a). If the phase sequence is reversed (RBY), then phase 'B' voltage lags phase 'R' voltage by $120°$ and phase 'Y' voltage lags phase 'B' voltage by $120°$. The phasor diagram for this case is shown in Fig. 2.13(b). In this case, the phase sequence is said to be negative.

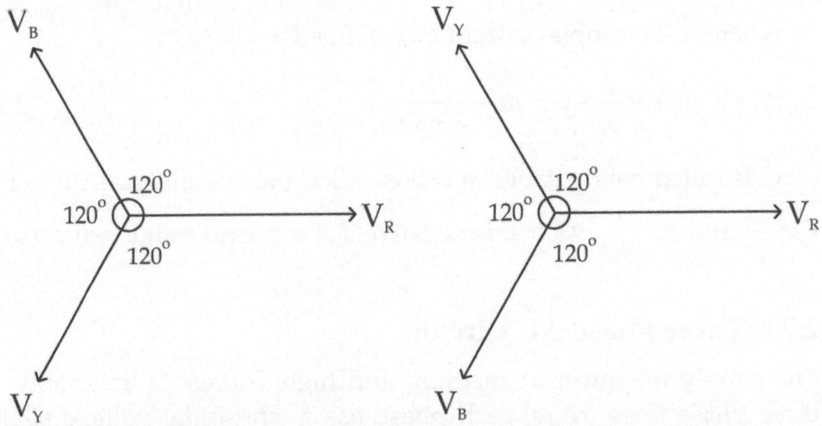

(a) Positive (phase) sequence **(b)** Negative (phase) sequence

Fig. 2.13 Phasor diagram of three phase voltages

The voltage across any two phases RY, YB and BR are also balanced and phase displaced by $120°$. However the magnitude of $V_{RY} = V_R - V_Y$ is $\sqrt{3}$ times the phase voltage magnitude (V_R or V_Y). Also V_{RY} leads V_R by $30°$ (for phase sequence RYB).

It is also possible to connect the voltage sources in a delta configuration (see Fig.2.14) with no common reference point.

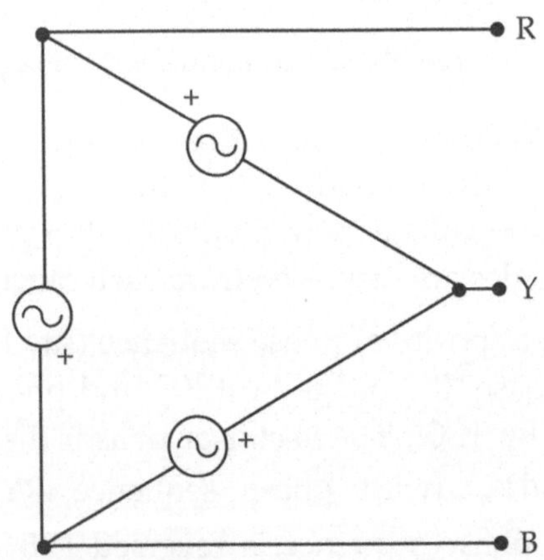

Fig. 2.14 Three phase delta connected voltage source

The loads can also be connected in star, also called as wye (with a common terminal N) or delta. If three single phase loads are connected across a balanced voltage supply, (which is also star connected) there is invariably a fourth wire connecting the two neutral (common) terminals for the supply and the load. If the load is balanced, it does not matter whether it is connected in star or delta. A balanced load is characterized by identical impedances connected to each phase (or across any two phases). The magnitude of the impedances for a balanced delta connected load are three times the magnitude of the impedances for a balanced star connected load, such that the line currents are identical in both cases.

A balanced load results in zero current flowing on the neutral wire. In addition, the three phase currents are balanced (of same magnitude but shifted in phases angles by $120°$ from one another). If the phase sequence of voltages is RYB, the current in phase 'Y' lags the phase 'R' current by $120°$. It is an accepted convention to label the phase sequence RYB (or ABC) as positive sequence and the sequence RBY (or ACB) as negative sequence. A positive or negative sequence balanced voltage supply results in balanced positive (or negative) sequence currents in the load if their impedances are identical in three phases. Irrespective of the sequence, we can show that the sum of the phase voltages (or currents) add to zero. However, in a 4 wire, three phase load, the current in the neutral wire is the sum of the three phase currents and is nonzero if the currents are unbalanced.

An important theoretical result given by C.L. Fortescue in 1918, states that any set of three phase sinusoidal voltages or currents can be expressed as a combination (sum) of three symmetrical components (i) positive sequence (ii) negative sequence and (iii) zero sequence. The last component has equal contribution to the three phases. It can be shown that the current flowing in the neutral wire is purely a zero sequence current that has equal contribution from three phases. If there is no neutral wire, there is no possibility for zero sequence currents to flow. A star connected load with isolated neutral point has no zero sequence current component. However, the neutral voltage can fluctuate. Only if a load is balanced (with equal impedance in each phase), and if the applied voltage has no zero sequence component, then the neutral point remains at the ground potential.

The utility of the concept of symmetrical components is that the currents flowing in a balanced three phase load are purely positive sequence if the applied voltages contain only positive sequence components. The generated voltages are purely positive sequence components under normal conditions and assuming the transmission network is symmetric (in the three phases), the loads are balanced, the currents flowing in the network contain only positive sequence components. This is an important fact and it implies that one can simplify the analysis of a three phase network by considering only an equivalent single phase network. The three phase powers and reactive powers are obtained as three times the values calculated for the equivalent single phase network.

2.8 Advantages of Three Phase Supply

There are several advantages of three phase supply as listed below.

1. Three phase generators and motors are more efficient than single phase generators and motors. For a given size, their ratings are 50% more than their single phase counterparts.
2. The sum of instantaneous powers in three phase symmetrical networks excited by pure positive sequence voltages is a constant = $3VI \cos \phi$, where V and I are phase voltage and current respectively (in any phase) and $\cos \phi$ is the power factor. There is no alternating component at twice the system frequency. This is due to the fact that sum of three alternating components produced by three phases is zero.

The implications of constant instantaneous power is that the mechanical stresses on the rotors of generators and motor are minimized. The alternating component of instantaneous power results in alternating component in the torque on the rotor which in turn results in mechanical stresses on the rotor and the shaft. It is to be noted that any switching action (including short circuits) in the vicinity of a generator can produce alternating components in the torque that decay with time. However, such incidents are few and far between.

Three phase networks do not require return conductor (unlike single phase networks). This is due to the fact that the sum of three phase currents add to zero at each instant.

The minimum number of phases having the property of not requiring return conductor is three. Increasing the number of phases beyond three results in unnecessary complexity without any significant advantage.

The induction motors, used extensively in industry are mostly three phase motors. Single phase motors, mainly used in residential loads (fans, for example) have no starting torques unlike three phase motors. Special arrangements have to be provided in single phase motors for starting.

2.9 Application of Circuit Analysis for a Transmission Line

Consider a 3 phase transmission line carrying power from location A to location B. in steady state, the line can be represented by an equivalent circuit (per phase) as shown in Fig. 2.15 (a). This shows the line can be represented by a series impedance $R + jX$ and two shunt admittances Y/2 (Y= jωC), C is the total shunt capacitance of the line, per phase) connected at A and B. Assuming the power and current flows from A to B, the voltage drop from the sending end and the receiving end is given by $(R + jX)\ I_L$. There is no loss of generality if V_B is assumed to be the reference phasor. If we assume the current I_B is in phase with the V_B and neglect the effect of line capacitances (which is valid for moderate line lengths), the magnitude of current in the line I_L ($\simeq I_B$) is given by (see the phasor diagram (2.15(b)).

$$I_L = \frac{V_A sin\delta}{X},\ X = 2\pi fL$$

(Note R is much less than X). L is the total series inductance of the line per phase.

In general, the current and the power flow in a line are determined by the voltage drop across the line. In AC lines, the voltage drop is mainly due to the phase difference (δ) between the voltages at the two ends. The magnitudes of the bus voltages have little effect on the current flow. Even if the magnitudes of V_A and V_B are equal, the current can flow due to the phase difference. (It is to be noted that we are denoting a phasor and its magnitude by the same symbol to simplify the notation).

In contrast, in DC lines, the current flow is the result of the difference in the magnitudes of the two bus voltages. Since line resistance is small, the current has to be regulated by automatic control.

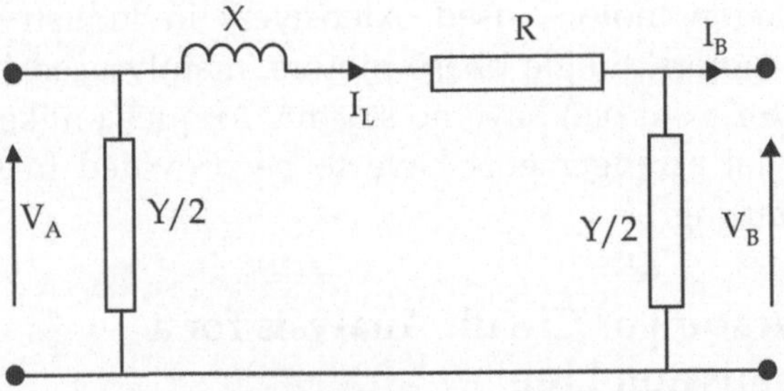

(a) Equivalent circuit

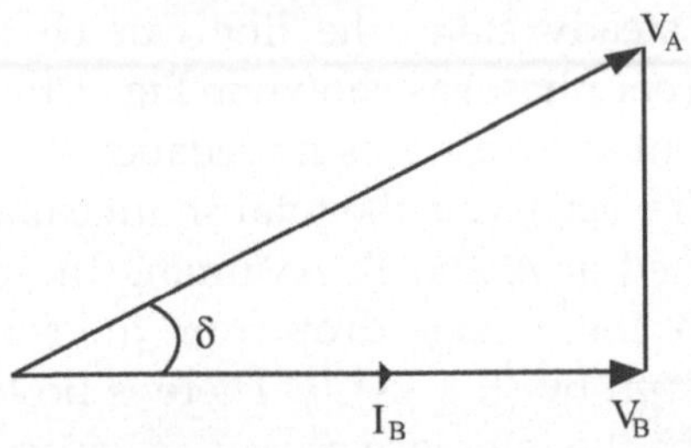

(b) Phasor diagram

Fig. 2.15 A transmission line representation

CHAPTER 3

Generators, Motors and Transformers

"There has been created in the generator something which exists nowhere else in nature."

Richard Feynman

3.1 Introduction

The power stations using fossil fuel, hydro or nuclear energy have electrical generators driven by either steam turbines or hydraulic turbines. The highest rating of generators operating in the system is around 1300 MW (driven by steam turbines). The largest generators driven by hydro-turbines is around 800 MW. The constraint on the size is primarily due to the bottlenecks in the transportation to the site where the power station is situated. To reduce the size of the generators, superconducting field windings were considered over 40 years ago, but were not implemented due to economic and reliability issues(It is to be noted that reliability is a major issue in the introduction of a new technology in power systems). As mentioned in Chapter 1, the primary objective in electricity supply is to provide electrical energy at the lowest cost while maintaining the desired level of reliability. Adding technological complexity, particularly when a new technology is being introduced can affect reliability. Hence, power engineers tend to be conservative in their approach in evaluating new (unproven) technologies.

The generators essentially convert energy from the mechanical form (kinetic energy in the turbines) to electrical energy. As mentioned in chapter 1, the generators are AC generators producing energy in three phases at medium voltages (below 30 kV). The generators are typically called as synchronous generators as they run at constant (synchronous) speeds. In thermal power stations (where steam turbines) are used, the generators run at 3000 or 1500 RPM (Revolutions Per Minute) when the frequency is 50 Hz. Actually, the speed (N) in RPM can be obtained by the simple relationship,

$$N = 60F/P \qquad \text{.....(3.1)}$$

where F is the frequency in Hz, P is the number of pole pairs. (Note that generators require magnetic field produced by an electromagnet with one or more pole pairs (North and South poles). This shows the highest speed of the generator corresponds to P =1. The generators in thermal stations have either one or two pole pairs. In contrast, the generators in hydro-stations have larger number of pole pairs (up to 10).

The electrical power generated at medium voltages has to be transformed to high voltages by using step up transformers. The three phase outputs of the AC synchronous generators are directly connected to the generator transformer to raise the voltage to the transmission level- 110 kV and higher. In India, maximum voltage level is 400 kV although 800 kV transmission has been introduced. 400 to 800 kV transmission lines are labeled as Extra High Voltage (EHV) lines. It is to be noted that the voltage level refers to the rated voltage between any two phases of a three phase line. In U.S.A., EHV lines are either 500 kV or 765 kV (maximum) lines. Generally, higher voltages are obtained by using autotransformers (for example, 800 kV is obtained from 400 kV level by an autotransformer) that will be described later in this chapter.

3.2 Principles of Electric Machines

The generators and motors are electrical machines that convert mechanical (kinetic) energy to electrical form and vice versa. This is termed as electromechanical energy conversion. In principle there is no difference between a generator and a motor. The same machine can operate as a generator or motor depending on whether mechanical energy is input or electrical energy is input. As mentioned in chapter 1, DC generators (that produce DC output) were replaced by AC generators although they may be still employed to excite the field windings of synchronous generators. There are basically two types of AC generators- synchronous generators (that have to operate at the same electrical frequency thereby implying constant speeds defined by equation (3.1)). These are predominantly used in power stations. The other major type is induction generators that are primarily used in wind power stations (driven by wind turbines). Induction motors are more popular in industrial applications and have replaced DC motors that were used previously.

Generators and motors have a stationary part called 'stator' and a rotating part called as 'rotor'. One of them produces magnetic field due to electrical excitation from windings supplied by DC or AC source. The other part also has windings or electrical circuits that allow flow of currents due to induced voltages. In synchronous generators, the magnetic field is produced by a DC source that is connected to the rotor winding (called as field winding) and the stator has three windings (called as armature windings) corresponding to the three phases in which voltages are induced (because of rotation of the rotor carrying the field winding).

The basic principles governing the operation of generators, motors and transformers are based on the interactions between the electric and magnetic fields. In 1819, a Danish physicist, Oersted found that a current carrying wire caused a nearby compass to deflect. This showed that electric currents have magnetic fields surrounding the conductors carrying the current. Subsequently, a year later, a French physicist, Ampere showed that the magnetic field produced is proportional to the current.

In 1831, Michael Faraday demonstrated in London that a changing magnetic field could produce an electric current. The discoveries of Oersted and Faraday show that (a) electricity can produce magnetism and (b) magnetism can produce electricity.

Faraday's law states that, the EMF (Electro-Motive Force) induced in a coil is given by

$$E = -\left(d\psi/dt\right) \quad \text{.....(3.2)}$$

where

E = total emf in Volts

ψ = total magnetic flux in Webers linking the coil

The negative sign in the above equation reflects the fact (stated as Lenz's law) that the current in the coil resulting from the induced emf is in a direction such that it opposes the change in the magnetic flux. If the coil has N turns, then the total flux refers to the flux linking all the turns. If the flux linking each turn is same and equal to φ, then $\psi = N\varphi$.

To distinguish between ψ and φ, ψ is referred as flux linkages and measured as Weber turns, although there is no essential difference in their dimensions (Note that a dimension defines some physical characteristic such as length, mass time, speed and force). Appendix A describes the dimensions and units of the various quantities used in physics. Weber is the unit of magnetic flux in International System (SI) of units used extensively now.

There are 3 ways which the total flux linking a coil can change. These are:

(i) If the flux is constant (not varying with time), the coil may move relative to the flux.

(ii) If the coil is stationary, the magnet producing the flux can move relative to the coil.

(iii) Both magnet and the coil are stationary, but the flux may vary with time. For example, in an electromagnet, flux is produced by the current in the field winding. If the current is changing with time, this will result in a change of flux linkages of the stationary coil in the vicinity.

Actually, there are no significant differences between the first two methods of inducing emf in a coil. For example, in a synchronous generator, DC field winding can be located in the stator and the three phase armature windings located in the rotor. However, practical considerations dictate the location of the DC field winding in the rotor. The terminals of the field winding are connected to the two slip rings that are rotating with the rotor. The current from the DC voltage source flows through the stationary brushes that make contact with the slip rings. If the armature windings are on the rotor, three slip rings are required to connect the three phase windings. The voltages across any two slip rings and the currents flowing through them will be much higher compared to the field voltage and current. For example, a 1000 MVA, 25 kV generator can supply 23000 A of rated current which the slip rings have to carry with 25 kV across any two slip rings if the windings are on the rotor. If a three phase fault occurs close to the generator, the fault currents can exceed 75000 A. In contrast the rated DC field voltage is 500 V and the rated DC current is about 6000 A.

The induced emf in cases (i) and (ii) is termed as rotational emf as the change in the flux linking the armature windings is caused by the rotation of the rotor. The emf induced when the two windings are stationary (case (iii)), is termed as transformer emf. In general, both types of emf are present in rotating electric machines. In contrast, transformers are stationary equipment whose operating principle is based on the induction of emf due to the time variation in the flux linkages of a stationary (secondary) winding caused by the time varying magnetic field produced by the exciting (primary) winding. The winding that supplies power to the magnetically coupled (secondary) windings is called as the primary winding. A transformer can be single phase or three phase. Although normally, a single-phase transformer has two windings, it is possible to have more than two windings. Also, it is possible to construct a three phase transformer with the help of three single-phase transformers. It is obvious that if a DC current flows in the primary winding, it will not induce an emf in the secondary winding. Thus, transformers can operate only with AC voltages for voltage transformation.

3.3 Rotating Electric Machines

EMF and Torque Production

Before we look into the working of different rotating electric machines, let us consider how an emf is induced in an armature coil by the relative motion of the coil and the electro-magnet. In a rotating electric machine there are two parts-stator and the rotor which are separated by an annular air-gap. The gap may be uniform around the circumference of the stator/rotor(in the case of a cylindrical rotor machine) or non-uniform in the case of salient-pole machines. In a DC generator, the DC field in the air-gap is caused by the DC field winding on the poles mounted on the stator, whereas in synchronous generators, the field and hence the salient poles are on the rotor. However, synchronous generators driven by high speed steam turbines, have cylindrical rotors with conductors of the field winding located in slots on the rotor surface.

A single coil with N turns has 2N conductors that experience relative motion with reference to a magnetic field having flux-density (B) in the air-gap. (See Fig.3.1).In general, the flux density is a function of both the position along the rotor circumference and time. The emf (e) induced in a single conductor of length (l) meters in a radial field is given by the following equation.

$$e = Blu \text{ Volts} \qquad \text{.....(3.3)}$$

where u is the velocity of the conductor in meters/second (in a direction at right angles to the field). The velocity is also equal to the product of the speed of rotation (in radians per second) and the radius of the rotor. The flux density is measured in Tesla (Webers per square meter). Note the earlier unit of flux density is Gauss (Webers per square centimeter). Thus 1 Tesla = 10,000 Gauss.

The emf induced in the conductor results in current (i) depending on the impedance of the circuit. The current interacts with the flux to produce a force (F) given by the equation,

$$F = Bli \text{ Newtons} \qquad \text{.....(3.4)}$$

where i is in Amperes and B is in Tesla.

Note that the force multiplied by the velocity is equal to the emf multiplied by the current. Both have the dimension of power and measured in Watts. Fig. 3.1 shows that the force acting on the two conductors (of one turn of a coil) are in opposite directions and produce a torque on the rotor which opposes the rotation of the rotor.

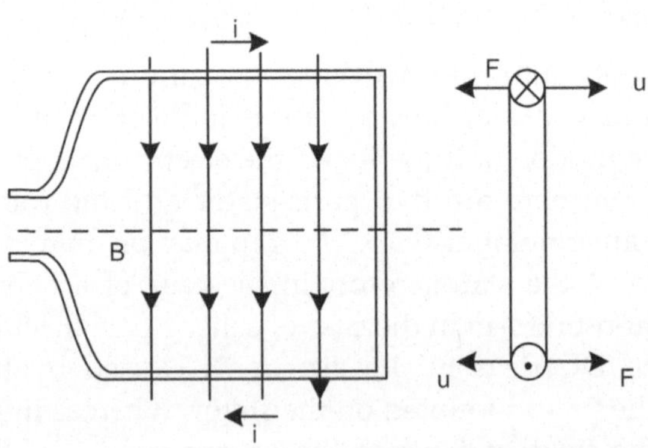

Fig. 3.1 A single coil in relative motion with respect to a magnetic field

This essentially implies that the electric power supplied by the generator is identical to the power input from the turbine driving the generator rotor. This assumes that there are no losses. However, in practice, there are mechanical losses caused by friction (at the bearings) and windage. Also, the output from the generator is less than the product of emf and the current as there are losses in the magnetic core and losses due to the resistances of the conductors.

The current flowing in the conductors can also be due to the applied voltage from an external source. In this case, the machine acts as a motor driving a mechanical load. In steady-state, the power supplied by the external source is equal to the power required to drive the motor against a load torque. In a motor, the electrical torque is in the same direction as that of the rotation whereas in a generator, the electrical torque is in the opposite direction.

Synchronous Machines

There are two types of synchronous machines cylindrical rotor and salient-pole rotor. In both types, the field winding is on the rotor. The cylindrical rotor is used high speed machines driven by steam turbines, whereas salient-pole rotors is used in lower speed generators driven by hydro-turbines. These two types are shown in Fig. 3.2. The salient pole generators operate at speeds between 80 and 600 RPM. They have very large diameters to accommodate the poles. The design of the rotor has to take into account the possibility of over-speeds (exceeding the rated speed) caused by the sudden loss of load. The rotor must be able to withstand the centrifugal forces during the over-speed conditions.

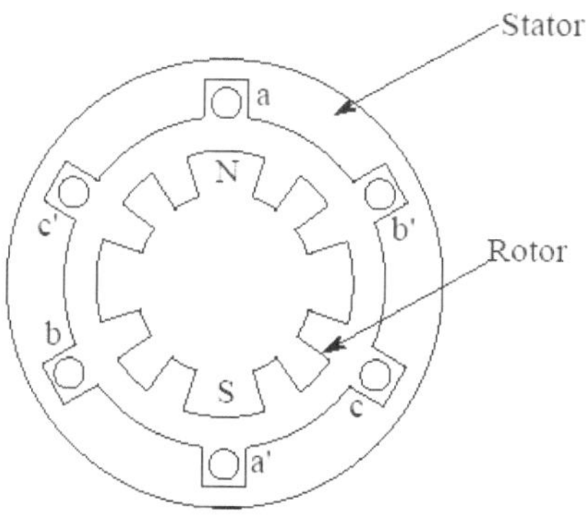

(a) Cylindrical rotor

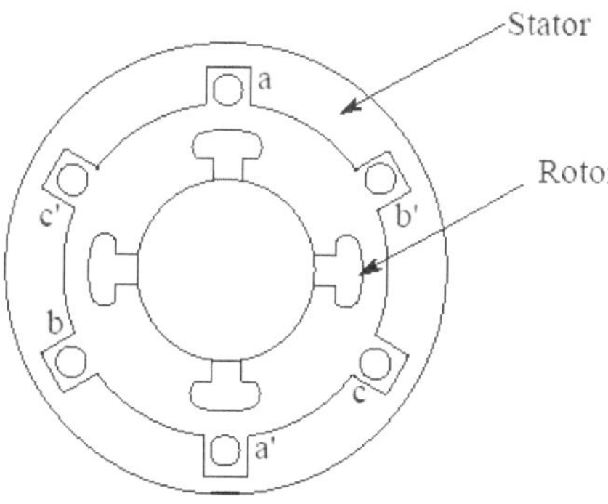

(b) Salient pole rotor

Fig. 3.2 Synchronous machine

The laminated poles are used in salient-pole machines. They are made of thin steel plates clamped between heavy end-plates and secured by rivets. The need for lamination arises due to the eddy currents flowing in the conducting ferromagnetic material in the poles. The eddy currents are caused by the emf induced in the magnetic core due to the changes (pulsations) in the flux as the rotor moves. The thin steel plates that are insulated from each other ensure lower eddy current losses in the magnetic core of the poles. The field winding is wound over the poles and is made of rectangular sections of copper conductors. To provide a damping torque against the rotor oscillations, coppers rods in the pole shoes are provided which are brazed to copper end bars. (The field winding and damper rods are not shown in the figure). Sometimes, the damper bars in the poles are connected together using inter-polar cage connections (as in the case of squirrel-cage induction motors). The salient–pole generators may have either horizontal or verticals shafts. The bearings for the vertical shaft generators must be capable of carrying the weight of the rotor, turbine runner and the hydraulic thrust.

The turbo alternators (synchronous generators with 2 or 4 poles and driven by steam turbines) have cylindrical rotors generally made of solid forgings of alloy steel. The field windings are placed in slots that occupy about two-thirds of the rotor pole pitch. Since copper windings are subjected to heat and centrifugal forces (that compress them against the slot wedges) they tend to undergo plastic deformation which results on contraction of the conductor when it is cooled. This phenomenon of copper shortening is overcome by the use of silver bearing copper that has higher yield point.

The stator windings are placed in slots which are distributed uniformly on the stator circumference near the air gap. (Note that Fig 3.2 shows only one stator coil per phase to simplify the concept). The number of slots for a three phase machines must be multiples of 3. The stator windings are either single -layer concentric or two-layer short -pitched windings to eliminate 5th and 7th harmonics from open-circuit emf.

The stator windings of large generators are directly cooled by circulating water either through hollow copper strands (of the conductor) or a combination of solid and hollow strands and other techniques for better control of heat dissipation. The rotors are cooled by hydrogen gas which has $1/14^{th}$ density of air (thereby reducing the windage loss and noise). Hydrogen has 14 times the specific heat; 7 times the thermal conductivity compared to air. This results in the reduction of temperature gradient (which also reduces the insulation corona). Hydrogen will not support combustion as long as the hydrogen/air mixture exceeds 3/1. Hydrogen is supplied above atmospheric pressure. The radial flow cooling is considered effective for 4 pole generators.

Performance of Synchronous Machines

The power output of a generator can be controlled by the control of input power. In a thermal power station, the chemical energy is converted into heat energy which is subsequently converted kinetic energy by the steam or gas turbines. The overall efficiency of energy conversion to the electrical energy is less than 40%. The Combined Cycle Gas Turbine (CCGT) has an overall conversion efficiency of about 50-60%. Starting a coal fired generating unit and bringing it to full load may take hours and even raising the power level in an operating unit may take several minutes. Thus, a large fossil-fuel fired or nuclear unit is normally operated as a base-loaded unit with constant power output.

The reactive power output of a generator is controlled by adjusting the field current that is a primary source for the magneto-motive force (MMF). The net MMF is also affected by the armature currents. When the AC current lags the generated voltage, the MMF produced by the armature currents opposes the MMF due to the field current. On the other hand, MMF of armature currents (leading the voltage) aids the MMF due to the field current. The effect of the armature currents on the net flux in the air gap and consequently on the generated voltages is termed as 'armature reaction'. The net flux in the air-gap of the machine is determined by the net MMF and the reluctance of the magnetic circuit in the machine. Note that magnetic circuit is analogous to electrical circuit with MMF analogous to EMF, flux analogous to the current and reluctance analogous too the resistance. However, the product of MMF and the magnetic flux has the dimension of energy rather than power. The path for the magnetic flux in the machine consists of ferro-magnetic material in the stator and the rotor in series with the air-gap.

A major difference between the electrical circuit and the magnetic circuit is that the magnetic flux conductors are poor in comparison with electrical conductors. The electrical conductivity of copper or aluminum is of the order of 10^7 mho per meter as compared to a porcelain insulator having a conductivity of about 10^{-14} mho/meter.

Air is a good insulator and thus conductors in a transmission line do not require any insulation (they are insulated from the tower by porcelain insulators). In contrast, the relative permeability of the best ferromagnetic material is 10^6. The relative permeability is the ratio of permeability of the material to that of free space, which is practically that of air. What is more important is that the relative permeability is not constant, but is a function of magnetic field (defined as MMF/meter). The flux density B, expressed as flux per unit area) increases initially in a linear fashion, there is saturation beyond a certain value of magnetic field (denoted by H). The variation of 'B' with 'H' is shown in Fig. 3.3. The implications of reduction in the relative permeability is that the magnetic flux has a tendency to stray from designated paths and pass through paths where it does not contribute to the generation of EMF or torque in the machine. The flux straying from the designated path is called as leakage flux. The design of the magnetic circuit should ensure minimization of the leakage flux.

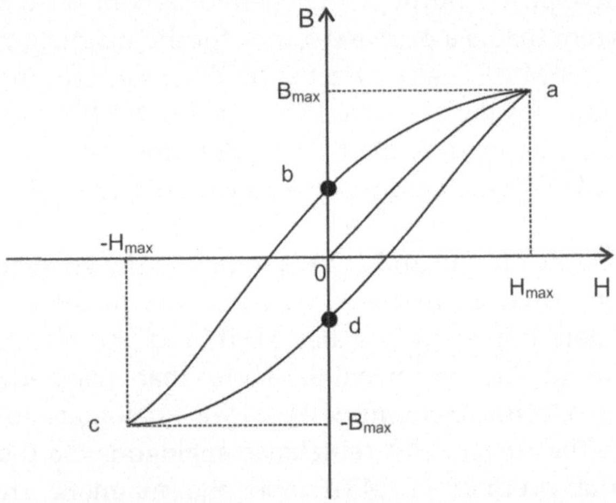

Fig. 3.3 B-H curve showing hysteresis

Setting up of DC magnetic flux requires energy, however no energy is required to maintain it. In contrast, pulsations of magnetic flux set up by AC currents result in power loss (called as hysteresis loss). In Fig 3.3, we have assumed that the initial flux density is zero when $H = 0$. The maximum flux density (B_{max}) corresponds to $H = H_{max}$. However, when the current is reduced to zero ($H = 0$), the flux does not become zero and there is remanent flux (corresponding to point 'b'). As the current is reversed and reaches a negative maximum value ($-H_{max}$), the B-H curve traces the path 'bc'. As the current is reduced to zero and again reaches positive maximum, the B-H curve traces the path 'cda'. The closed curve 'abcda' represents the energy loss caused in the magnetic material per unit volume in one cycle of the current. The hysteresis loss (in watts) is proportional to the frequency of the magnetizing current. It is also a function of the material and the maximum flux density (determined by the magnitude and frequency of applied voltage, area of the magnetic core and the number of turns of the coil producing the magnetic field).

In addition to the hysteresis loss, the (magnetic) core losses also include eddy current losses cause by induced currents flowing in the core (as it is not an electric insulator). These losses are reduced by using laminated core and choosing a material of lower conductivity. For a given material, the eddy current losses are proportional to the square of the frequency and maximum flux density.

Synchronization of a Generator

Before a synchronous generator is connected to the grid it is necessary to check

(i) the generated voltage magnitude is approximately equal to the system voltage (measured at the generator bus)

(ii) the speed of rotation is close to the rated speed

(iii) the phase difference between the generator and system voltage is close to zero

A simple means of checking the above conditions is by connecting a lamp across the switch (breaker) in each phase (two being cross connected while the third connected directly across the switch). When the switch is open, and the generator frequency is slightly different from the system frequency, the three lamps slowly brighten and darken in cyclic fashion. The main switch can be closed safely when the cross connector lamps remain steady and voltage across the switch is zero. As a matter of fact, the generator gets synchronized after the switch is closed and the generator frequency matches exactly with the system frequency in steady state (when the power and reactive power outputs remain steady). Closing the switch at wrong instant can result in large currents and pulsating torques that could damage the shafts. A synchroscope is used to ensure the switch is closed at proper instant of time when the three conditions mentioned earlier are met.

The operating region (in P-Q variables) is bound by the following constraints:

(a) The maximum power output is determined by the limits on turbine power

(b) The reactive power supply is limited by the upper limit on the field current as well as the upper limit on the armature current

(c) The reactive power absorption (operation at leading power factor) is limited by stability considerations

The rated power factor is typically 0.9.

Need for Fast Voltage Regulation

Prior to 1950s the voltage regulation was manual. In fifty's rotating amplifier (amplidyse) and magnetic amplifiers were applied.

The effect of armature reaction (caused by the armature currents) becomes predominant as the generator sizes (ratings) increase. Optimal design of generators has resulted in reducing an important parameter (Short Circuit Ratio). The SCR is defined as

$$\text{SCR} = \frac{\text{Field current required to produce rated voltage on no load (open circuit)}}{\text{Field current required to circulate rated current in the armature when the terminals are short circuited}}$$

The SCR has reduced to 0.5 from values over 1.0 in the earlier days.

A generator cannot stably supply rated power unless the voltage is regulated fast using Automatic Voltage Regulator (AVR) and a fast acting field exciter. Previously DC generators driven by the turbine on the same shaft as the generator were used. In the sixties, static excitation system was introduced. Here, the voltage and current transformers connected to the generator terminals supply a static power converter (employing power semiconductor devices) to supply the field winding. The output voltage of the converter can be controlled rapidly by adjusting the switching instants of turning on the thyristor switches. The controller is an electronic amplifier with a high gain. Comparing the voltage reference signal with a signal derived from the measurement of the terminal voltage, the error (signal) drives the amplifier whose output is the desired field voltage required to regulate the generator terminal voltage. Fast regulation of the terminal voltage compensates the effects of the armature reaction.

DC Machine

A DC machine is similar in construction to the salient-pole synchronous machine except that the field winding is on stator. The armature winding is on the rotor which has AC voltages induced in the conductors that are moving alternately under the north and south poles. To convert AC voltage to DC, the individual coils are connected to commutator with a number of segments that are insulated from each other. The stationary brushes making contact with the rotating commutator mounted on the rotor shaft result in the rectification of AC voltage to DC. The commutator may be viewed as a rotary switch which ensures that the voltage across the brushes is DC although the induced voltage in the rotor coils is AC. The position of the brushes is usually in the inter-polar region where the magnetic field produced by the field winding is negligible. However, armature reaction does affect the net magnetic field and this may require shifting the brushes. The magnitude of the current that can pass between the commutator and brushes without causing serious sparking (and consequent damage) due to high current density in the contact area, does limit the rate of increase in the power output of the DC machine.

The voltage rating of a DC machine is affected by the potential difference that can exist between adjacent commutator segments - a typical value is 50V/cm of commutator periphery. The brushes are made of carbon while the commutator segments are made of copper. Increased electric field on the commutator can cause flashover between the segments due to carbon dust from the brushes and ionized air in the neighbourhood.

As mentioned earlier, DC generators have long been replaced by AC generators. DC generators were used to supply the field current of the AC generator. For reliability, the rotor of the DC generator is also mounted on the same shaft driving the AC generator. The DC generator field winding may be (i) separately excited or (ii) shunt excited (supplied by the DC generator). In the former case a pilot exciter (smaller DC generator) used to supply the field current of the main exciter.

DC motors have been used extensively in the past, in industrial applications because of its excellent torque-speed characteristics. The torque produced by a DC motor is proportional to the product of armature current and the field flux. The armature current increases in response to the increase in the required (mechanical) torque. For a given field current, the speed is practically constant (with a slight droop as the voltage drop increases with the load). The emf induced in the armature is directly proportional to the magnetic field and the speed of the motor. It is also determined by the applied voltage. Thus, weakening the field by reducing the field current results in increase in speed.

A DC motor can have the field winding connected in series with the armature. This arrangement results in a large torque at starting which is required in some applications. For low load conditions, the speed of the motor tends to rise as the field current is reduced.

If a DC series motor is supplied by AC source, it would continue to run as both the armature current and the field current reverse as the voltage is reversed and the torque would be unidirectional. Motors designed for operation both for DC and AC supply are called as universal motors. They are designed for power output less than one kilowatt and operate at speeds of 4000 to 10000 RPM. The advantage of a series motor is that it cannot be overloaded as the product of torque and the speed is practically constant.

The DC exciter used in power stations to supply the field current in the synchronous generator can be brushless and use rotating power diodes to rectify the AC voltage induced in the armature windings. This is more rugged, faster in response and requires less maintenance than commutator and brushes.

Induction Machines

In an induction machine, three phase stator winding functions both as exciting (producing the magnetic field) and armature winding. The rotor is either (a) squirrel cage or (b) wound rotor. The squirrel cage rotor is made of a number of conductor bars connected to end rings. A wound rotor has three phase winding which can be brought out using slip rings. The windings are short circuited directly or in series with variable resistance in each phase.

The magnetic field produced by the symmetrically placed three phase stator windings excited by balanced currents results in a rotating magnetic field in the air gap. This induces a voltage in the conductors of the rotor and currents flow that interact with the magnetic field in the air gap to produce a torque in the same direction as the rotating magnetic field. The rotor moves in the same direction as the rotating magnetic field. The frequency of the voltage induced in the rotor is the difference in the speeds of the rotating field and the rotor. The magnitude of the voltage induced is proportional to the frequency (also called as the slip frequency) of the induced voltage. It is obvious that when the slip frequency is zero, then no currents flow in the rotor windings and the torque produced is zero. Thus, an induction motor can never run at synchronous speed. It will always be less than the synchronous speed. Note that for a 50 Hz, 4 pole motor, the synchronous speed is 1500 RPM.

Actually, the torque produced in the motor increases from zero (at synchronous speed) to a maximum value (called as the pull-out torque) as the rotor speed is reduced. The slip at which the torque is maximum depends on the resistance of the rotor winding. While the squirrel cage motor has no control over the torque at a given speed, the wound rotor induction motor can be controlled to adjust the torque by inserting resistance in the rotor circuits by connecting it externally through the slip rings. However, from the considerations of efficiency and power factor, it is desirable to operate at low values of slip. Thus, squirrel cage induction machines are called as fixed speed motors as the variation in the slip over the entire range of load torque may be 5% or less. Although induction motors are rugged, they were not considered suitable for adjustable speed drives. However, in recent times, the power electronic control of induction machines has made them as flexible as DC machines.

Induction machines operate as generators when they are driven by wind turbines to run at speeds higher than synchronous speeds. Wind power systems use mainly induction generators - squirrel cage machines as well as doubly fed induction generators (DFIG). In the latter, the energy that would have been wasted in the rotor (external) resistances is fed back to the supply using a diode rectifier (or a VAC) and an inverter.

3.4 Transformer

A single phase transformer with two windings wound on the two legs of a magnetic core is schematically shown in Fig. 3.4. A three phase transformer can be constructed as a single unit (with 3 limbs) or connecting three single phase transformers in either star connection or delta connection. Generally, a transformer has two sets of windings - primary and secondary. In some case, a single phase transformer can have 3 windings (the third winding is called as tertiary and may be used for tapping power).

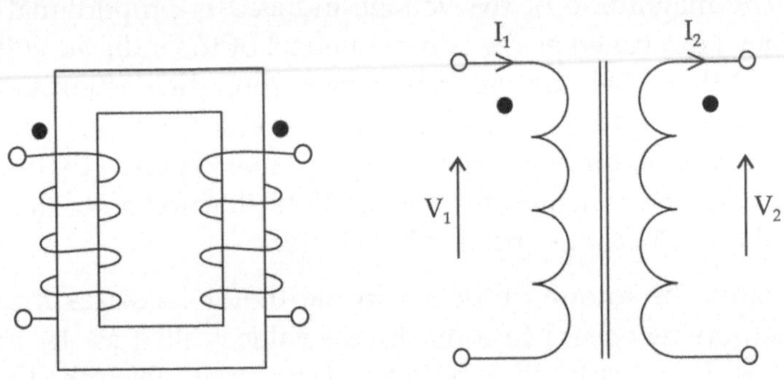

(a) Structure **(b)** Schematic diagram

Fig. 3.4 A two winding single phase transformer

The polarities (shown by dots) on the windings indicate how the MMFs created by the currents flowing in the windings contribute to the magnetic flux in the core. If the currents in all the windings either enter or leave the windings at the terminals with the polarity markings, the MMFs add up. In the two winding transformer shown in Fig. 3.4, the current in the secondary winding, I_2 is in a direction that opposes the magnetic flux caused by the current I_1 in the primary winding (by Lenz's law).

A transformer is a stationary equipment with no rotating parts. The major components of a transformer are:

(a) Magnetic circuits comprising limbs, yokes and clamping structures

(b) Electric circuits comprising windings, formers, insulation and bracing

(c) Terminals, tappings and tapping switches, terminal insulators and leads

(d) Tank, oil, cooling devices, dryers etc

To minimize the length of the conductors used, the coils are made circular which are wound around stepped core (see Fig 3.5).

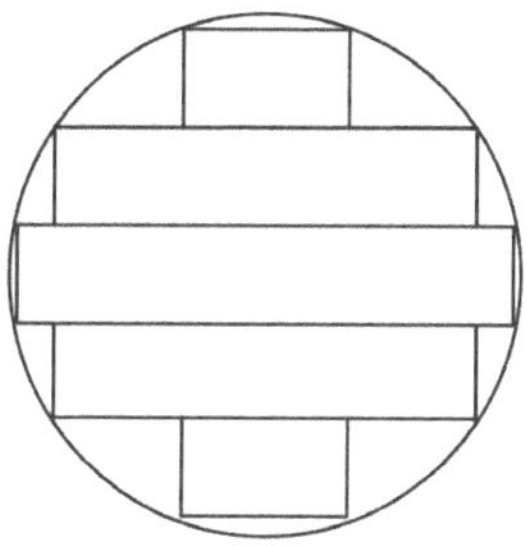

Fig. 3.5 Stepped core

For a single phase transformer (whose schematic is shown in Fig. 3.4), the voltage across the secondary (V_2) is related to the voltage across the primary winding by the turns ratio. The following equation applies

$$\frac{V_2}{V_1} = \frac{N_2}{N_1} = \frac{I_1}{I_2}$$

The current relationship follows from the fact that $V_1 I_1 = V_2 I_2$. The above expression assumes that leakage flux is neglected and the magnetizing current is very small. These assumptions are generally valid for power transformers used in generating stations and load centres. However, leakage flux and the resultant leakage reactance is not small. At full load, the voltage drop in the leakage reactance can be as large as 15% of the rated voltage. To regulate the voltage, On-Load Tap Changers (OLTC) are provided to adjust the turns ratio to maintain constant voltage.

The power losses in the transformer are due to (i) core losses (hysteresis and eddy current losses) and (ii) copper losses (due to winding resistances). These losses cause heat and to avoid damage to the insulation, it is necessary to provide heat dissipation by cooling. The heat developed in the cores and windings is passed on to the oil in which the transformer is immersed. The heat in the oil is dissipated through the tank walls with natural cooling. However, this may not be adequate for larger transformers and forced cooling is used by pumping oil to a refrigerator where air or water is used to convey the heat to the atmosphere.

Transformers in operation generate acoustic noise caused by the phenomenon of magnetostriction due to which, the laminations tend to vibrate by magnetic force. The noise is transmitted through the oil to the tank walls and result in a humming sound. The phenomenon of magnetostriction involves a very small elongation (with corresponding reduction of the cross-section) when sheet-steel strips are magnetized.

The noise is also due to the aerodynamic noise produced by cooling fans and the flow noise due to oil circulating pumps. However, the noise produced in the core at twice the system frequency (100 or 120 Hz) is the most troublesome.

Transformers used in distribution systems are designed to have smaller leakage reactance to reduce voltage variation with the load.

The efficiency of power transformers is more than 99%. For distribution transformers, the efficiency varies with the load - while the core losses are constant, the copper losses are proportional to the square of the load current. To reduce core losses, amorphous cores are used.

Auto Transformer

In an auto-transformer, the coils are shared by the two windings (see Fig. 3.6). It can also be viewed as a single tapped winding on a transformer core. The advantage of an auto-transformer is (a) reduction in cost and losses as the common winding carries a smaller current than a two winding transformer and (b) reduction in the leakage reactance that results in better voltage regulation. These advantages are predominant when the turns ratio is not high (typically ≤ 2). Autotransformers are used for

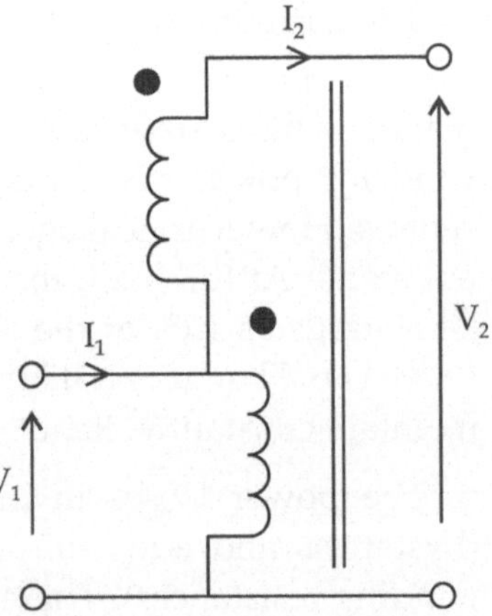

Fig. 3.6 An autotransformer

interconnecting transmission lines at different voltages and also raise the voltage level of a line (say from 400kV to 800kV) assuming that the line is designed for higher voltage.

The disadvantage of autotransformer is the lack of electrical isolation.

Condition Monitoring of Transformers

A transformer requires a long time to repair if the insulation of the windings is damaged. Hence, it is desirable to monitor the health of a transformer to protect it from destructive events. A typical transformer monitoring system has the following features:

1. Sensors for (a) combustible gases; (b) top oil temperature; (c) load current; (d) moisture in oil and (e) cooling system motor load current
2. Remote reporting of real-time data
3. Generation of alarms
4. Analysis of (a) winding hottest spot; (b) moisture in insulation; (c) insulation aging rate and cumulative aging and (d) cooling system efficiency
5. Sensors for partial discharge in bushings

3.5 Development of Electrical Technology

When Michael Faraday first made public his discovery that a changing magnetic flux produced an emf, he was asked "what is the use of it?" After all, he had found that a tiny current flowed in a wire when he moved it near a magnet. Of what possible use could that be? His answer was "What is the use of a newborn baby?"

However, his discovery led to tremendous practical applications starting with the generation and utilization of electrical energy. This was the beginning of electricity supply industry (ESI). There are motors that vary in size from fractional horse power (HP) to a few thousand HP depending on a variety of applications. There are also linear motors that are applied in transportation. Faraday's discovery is applicable in micro machines and actuators.

Richard Feynman, a theoretical physicist and a Nobel laureate, marvels at the beauty of Boulder dam and the hydroelectric power station there, converting potential energy in the stored water to electrical energy using penstock, water turbines and the AC generator. He says (Lectures on Physics) "The same phenomena that take the tremendous power of the river and spread it through the countryside, until a few drops of the river are running through the dentist's drill, come again into the building of extremely fine instruments, for the detection of incredibly small amounts of current, for transmission of voices, music and pictures, for computers, for automatic machines of fantastic precision. All this is possible because of carefully designed arrangements of copper and iron-efficiently created magnetic fields, blocks of rotating iron six feet in diameter whirling with clearances of 1/16 of an inch, careful proportions of copper for the optimum efficiency, strange shapes, all serving a purpose, like the curve of a dam. If some future archaeologist uncovers Boulder dam, we may guess that he would admire the beauty of its curves. But also the explorers from some great future civilizations will look at the generators and transformers and say: "Notice that every iron piece has a beautifully efficient shape. Think of the thought that you have gone into every piece of copper!"

Feynman goes on to say, "This is the power of engineering and careful design of our electrical technology. There has been created in the generator something which exists nowhere else in nature. It is true that there are forces of induction in other places. Certainly in some places around the sun and stars there are effects of electromagnetic induction. Perhaps also (though it is not certain) the magnetic field of the earth is maintained by an analog of an electric generator that operates on circulating currents in the interior of the earth. But nowhere have there been pieces put together with moving parts to generate electrical power as is done in the generator-with great efficiency and regularity. You may think that designing electric generators is no longer an interesting subject, that is a dead subject because they are all designed. Even if this were true, we can admire the wonderful accomplishment of a problem solved to near perfection. But there remain as many unfinished problems. Even generators and transformers are returning as problems. It is likely that the whole field of low temperatures and superconductors will soon be applied to the problem of electric power distribution. With a radically new factor in the problem, new optimum designs will have to be created. Power networks in the future may have little resemblance to those of today. Modern electrical technology began with Faraday's discoveries. The useless baby developed into a prodigy and changed the face of the earth in ways its proud father could never have imagined."

It is true that newer problems have cropped up with time. The design of electric machinery led to the development of adjustable speed drives with the discovery of power semiconductor devices such as thyristor and IGBT (Insulated Gate Bipolar Transistor). In addition to the pressing need for precise control of motors, there is also the imperative of energy conservation by reducing losses. High speed electric rail transportation utilizes magnetic levitation and superconducting magnets. The need to reduce the carbon foot print spurs the use of IGBTs in electronic ignition. The application of satellite communication and GPS technology in addition to robotic control, gives rise to driverless cars and guided highways.

In power transmission, power semiconductor devices and microprocessor control has given rise to UHV DC transmission that can carry 5000 to 6400 MWs of power over long distances while at the same time, back to back (BTB) links that are used to interconnect AC grids.

CHAPTER 4

Transmission and Distribution of Power

"We must all hang together, or assuredly we shall all hang separately."

B. Franklin

4.1 Introduction

In this chapter, we will consider the delivery of power generated in power stations to the customers who would use electrical energy at low or medium voltages. The generating stations may be located several hundreds of kilometers away from the load centres around urban areas where bulk power is distributed to factories (industrial loads), commercial establishments such as shops and offices (commercial loads) and homes (residential loads). These are three broad classifications of loads; however there are other categories of loads such as street lighting, agricultural (water pumping) etc. We will consider both AC and DC transmission and also briefly discuss the distribution of power at low or medium voltages. It is to be noted that the distinction between the transmission and distribution is somewhat artificial. However, there are important differences between the two networks apart from the differences in the voltages levels. Some large industrial and commercial customers receive power at intermediate voltage (sub-transmission) levels (33 kV to 115 kV).

As mentioned in Chapter 1, DC transmission at low voltages (voltage level of the DC generator) is not feasible as long distance transmission requires high or extra high voltages to minimise the power losses due to transmission line resistances. This is the major reason for the adoption of AC technology replacing the DC technology advocated by Thomas Edison who started the first power station in USA for street lighting. However, we will see later in this chapter, that high to extra-high voltage DC transmission (up to ±600 kV) is beneficial for long distance power transmission and has been used in several cases. For very long distance transmission (around 2000 km), China has gone in for the so called UHV (Ultra High Voltage) DC transmission at ±800 kVfor transmission of bulk power (6400 MW). India also has taken a decision of using UHV DC transmission for evacuation of hydro power from the North-East. However, HVDC power transmission requires the conversion of power from AC to DC at the sending end and from DC to AC at the receiving end by using high power electronic converters.

4.2 AC Transmission

To standardise the equipment used for AC power transmission, the voltage levels have been standardised. The rated voltages between any two phases (also called as line-to-line voltages) are (in kilo volts) as shown in Table 4.1

Table 4.1 Standard Voltage Levels

Nominal Voltage	132	220	275	345	400	500	750	1000	1150
Maximum Voltage	145	245	300	362	420	525	800	1050	1200

All these voltage levels are not used in India. The voltage levels used so far are 132 kV, 220 kV and 400 kV (nominal). At higher voltage levels, the line voltages are expressed in terms of the maximum operating voltages rather than the nominal values. Thus, 800 kV (maximum) voltage lines have been introduced recently in India whereas in USA the maximum voltage level is 765 kV corresponding to the nominal value of 750 kV. China is operating a test line at 1100 kV (maximum) while India is planning 1200 kV (maximum).

All long distance AC transmission use overhead wires (conductors) which are strung from towers by insulator strings to avoid contact with the metallic towers that are grounded. Only in urban areas, underground cables are used for short distances. The cost of the underground cable is much higher than overhead transmission. Although, cables (laid in trenches) are immune to weather fluctuations, the faults in cable transmission tend to be permanent (not self-clearing) and take more time to repair. In AC transmission, cables are not feasible for long distances for reasons that will be discussed later.

The transmission towers used are mainly self supporting lattice type. A well designed lattice structure makes the most efficient use of material which implies that it is light in weight and can be erected without the need for heavy equipment or access roads. However, many people think that lattice structures are ugly and that it needs a wider Right of Way (RoW). An alternative tower structure is guyed V or single column type, which tend to be taller.

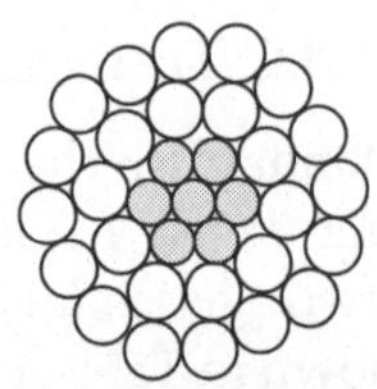

Fig. 4.1 ACSR conductor

The transmission line conductors are made of Aluminium. Although the conductivity of aluminium is smaller than that of copper, it is cheaper and has lighter weight. The most common conductor type is ACSR (Aluminium Conductor Steel Reinforced) which consists of layers of aluminium strands surrounding a central core of steel strands (see Fig. 4.1). Stranded conductors are easier to handle (more flexible) than solid conductors. Also the skin effect is minimized (skin effect refers to the tendency of the AC current in a solid conductor to crowd toward the edges or skin (rather than being uniformly distributed). The skin effect is due to Faraday's law and the fact that flux within the conductor is changing with time for AC currents. Note that there will be no skin effect with DC currents.

Other conductor types are - AAC (All Aluminium Conductor), AAAC (All Aluminium – Alloy Conductor) and Alumoweld (Aluminium-clad steel conductor).

EHV lines have sub-conductors in each phase, which are called as bundle conductors. There could be two, four or more sub-conductors in a bundle. Their use is dictated by the need to reduce the effects of corona discharge from the conductor surfaces, caused by break-down of air surrounding the conductor due to high concentration of electric fields at the conductor surface. Although the air breaks down at electric field of about 30 kV/cm (at atmospheric pressure and 20° C temperature), the roughness of the conductor surface can reduce it to half that value where there are surface irregularities. The corona is explained as partial discharges that occur when the electric field near the conductor causes ionization of the air surrounding the conductor. The free electrons in the air accelerate due to the energy provided by the field and dislodge electrons of atoms of nitrogen and oxygen in the air. The process involved in corona is quite complex. However, we can look at the effects which influence the line design. These are:

1. Corona Loss
2. Audible Noise (AN)
3. Radio Interference (RI)
4. Television Interference (TVI)
5. Ozone Formation and
6. Emission of visible light producing the corona glow

The losses occurring due to corona discharges in EHV lines vary from 1-20 kW/km in fair weather (with higher values corresponding to higher voltages); in foul weather, the losses can be as high as 300 kW/km. In contrast, the power loses (I^2R) in the conductor resistance (which gets converted to heat increasing the conductor temperature) can vary from 80 kW/km to 130 kW/km for a 400 km long line. I^2R losses do not vary too much with the weather (except for the fact that the conductor resistance increases slightly with increase in temperature). However, since rain fall is seasonal and location dependent, the estimate for the average corona losses (over a year) for a 400 kV line, 400 km long, varies from 2 kW/km to 100 kW/km. Thus, average corona losses are only about 10% of the average I^2R losses which also depend on the load factor. The load factor is the ratio of average load to the rated load. However, during foul weather, higher losses occur and this is a factor that affects the performance.

The RI is an important factor for EHV lines and is more significant than the losses. The Radio Interference occurs in the frequency range of 0.5 MHz to 1.6 MHz. In addition to RI, TV reception can also be affected in the 80-200 MHz range. Corona can also cause interference to power line carrier communication (30 kHz - 500 kHz) used for control and protection. For UHV transmission, the Audible Noise (AN) primarily determines the line design for corona effects. The maximum AN at the edge of RoW should be below 55dB.

4.3 Performance of AC Transmission Lines

Under normal conditions a transmission line carries power within the limits of the capability of the line. The actual power flow is a function of the injected powers at the generating stations, requirements of the power demand at the various sub-stations supplying the customers. In addition, the power flow is a function of the series impedances of the line as Kirchhoff's Voltage Law (KVL) dictates the current flow through various circuit elements in a mesh type network. In radial networks, the power flows in lines are determined by KCL and power injections at generator and load nodes (At a load node, the injection is treated as negative). Note that a transmission line can be viewed as a circuit element for analyzing the flow of currents.

The series impedance of a line is due to the resistance and inductance of the line (which results from the fact that magnetic fields are associated with the current flowing in the conductor). The flux density 'B' at a point at a distance 'x' from the conductor carrying current 'I' is given by

$$B = \frac{\mu_0 I}{2\pi x}, \quad \mu_0 = 4\pi \times 10^{-7}$$

Assuming there are no other conductors in the vicinity, the flux lines are circular and concentric with the conductor. μ_0 is the permeability of the air in Henry/meter.

A three phase line has both self and mutual inductances (a mutual inductance between phases 'R' and 'Y' is the result of flux linking Y (or R) phase due to current in R (or Y) phase. If the transmission line is symmetric, then all the three self inductances (phase R, Y and B) are equal and the three mutual inductances (between any two phases) are equal. If we assume only positive sequence currents (balanced in all three phases with sequence matching that of the voltage) flowing in the three phases, then the (positive sequence) inductance per phase is given by subtracting the mutual inductance from self inductance. Typically, the inductance of a 400 kV line per phase is one milliHenry per km. The inductance reduces as the voltage level of the line increases.

The (positive sequence) series impedance of the line is $(R + jX)$ where R is the resistance and X is the reactance $(2\pi fL)$. Note L is the inductance. Typically, the resistance is one tenth of the reactance or less. A line also has shunt capacitances. The positive sequence shunt capacitance per phase is about 11 nanoFarad (11×10^{-9} F) per km.

It is to be noted that the three parameters R, L and C are distributed in nature. The analysis of a line with distributed parameters can be quite complex. However, in AC circuits, in steady-state, we assume the voltages and currents to be sinusoidal and represented by phasors (see Chapter 2 for discussion of phasors). Both voltage and current phasors in the line vary continuously with distance from the sending end. The phase angle continues to increase uniformly with distance. When the length of the line is such that the phase difference between the two ends is 360^0, the line length is of one wave length. The wave length is inversely proportional to the frequency and for frequency of 50 Hz, the wave length is about 6000 km. AC transmission over 1000 km is not feasible unless special reactive compensation methods are used.

Control of reactive power and voltage are important considerations for designing a long distance AC transmission line. Due to the series reactance of the line, reactive power is consumed in the line. On the other hand, the shunt capacitance of the line acts as a source of reactive power. Since the reactive power consumed in the series reactance is proportional to square of the line current, whereas the reactive power generated by the shunt capacitance is proportional to the square of the line voltage (which needs to be maintained around the rated value), it is obvious that the net reactive power drawn by the line (from one or both ends) is negative at no load (power transferred is zero) and increases as the power transmitted is increased.

For a lossless line, there is a specific loading on the line, called as surge impedance loading (SIL) when the voltage remains constant at all points along the line. Also, the current and voltage phasors at all points in the line have identical phase angles and the power factor is unity. This implies that reactive powers supplied at both ends of the line are zero. This loading is an economic loading as not only the voltage profile is flat along the line, the reactive power requirements are nil. The surge impedance loading is given by

$$SIL = \frac{V^2}{Z_n}, Z_n = \sqrt{\frac{L}{C}}$$

where V is the rated (line to line) voltage, Z_n is the surge impedance which depends on the ratio of L (inductance) to the capacitance (C). L is expressed in Henry and C in Farad. The surge impedance varies between 200-400 ohms. The value decreases as the voltage level increases. For a 400 kV line, $Z_n \simeq 300$ ohms and SIL is about 500 MW.

It is to be noted that SIL is not a limit on the power flow in the line. The maximum limit of power flow is determined by the peak temperature permissible of the conductor (75° C), the ambient temperature and the power losses in the conductor. This is called as the thermal limit which is not fixed but varies with the season. In winter, with lower ambient temperatures, higher loadings are feasible. As a matter of fact, the dynamic rating of a line based on temperature sensors is a better option to make the best use of available transmission capacity. The increase in the peak temperature of the conductor results in increasing the sag of the conductor which can result in short circuits if there is vegetation below the line.

There are two other limits which can be decisive in determining the peak loading depending on the line length. For line lengths in the range of 100-300 km, the voltage drop across line determines the peak loading. However, above 300 km, the stability of the system becomes important criterion. The synchronous stability is a major aspect of system stability which involves ensuring that all the generators remain in synchronism even under large disturbances (such as faults in a line).

It is not sufficient to ensure the line loading does not cross any of the three limits mentioned earlier. It is also important to keep some margin to ensure security. Considering all these factors, the loadability of an AC line as a function of length of the line is shown in Fig. 4.2.

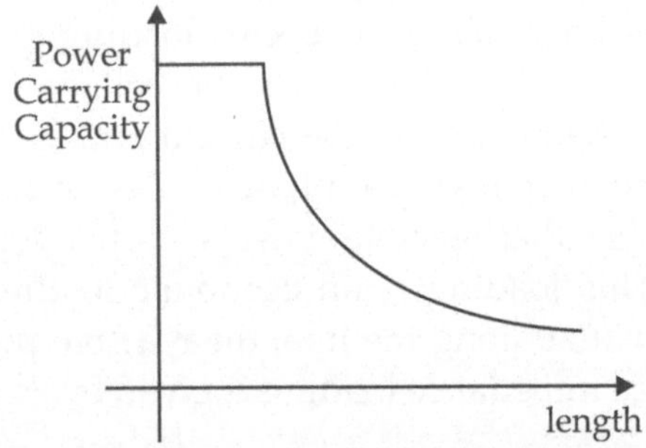

Fig. 4.2 Power transfer capability of an AC line as a function of length

Note that the power flow limit decreases as the line length increases due to the stability criterion and the variation is approximately in the inverse ratio. It should be noted that power flow in an AC line cannot be controlled or regulated unless special controllers using high power electronics are employed. These are known by the generic name FACTS (Flexible AC Transmission System) Controllers. The emerging technological innovations to optimize the working of electricity supply and minimize blackouts will be discussed in Chapter 7.

Although power flow in AC lines cannot be controlled easily, it can also be arranged that they do not require control. For example, consider a line that is connected between a power station and a load centre. If there are no generators at the load centre, the power flow in the AC line automatically adjusts to carry the power demanded at the load centre. This comes about because the power flow in an AC line is a function of the difference in the phase angles of the voltages at the sending and receiving end. There is no direct control of this angular difference. It changes automatically in response to the load demand. For a lossless line, the expression for the power flow in the line is given by

$$P = \frac{V_S V_R}{X} sin\delta, \delta = \delta_S - \delta_R$$

where δ is the angular difference between the sending and receiving end voltages (V_S and V_R). X is the total series reactance of the line. It is obvious that the theoretical maximum power that can be transmitted is when $\delta = 90^0$. Since the voltages at the two ends of the line have to be maintained close to the rated values, the power flow in the line for a specified value of the angle can be increased if the net total reactance is reduced by inserting a series capacitor whose reactance is considered to be negative and subtracts from the line reactance that cannot be altered. Although maximum power is obtained at $\delta = 90^0$, the operating value of the angle is not allowed to go above a certain limit which depends upon the stability margin desired. Since the maximum power is also affected by the impedances of the systems connected (at the two ends) of a transmission line, a thumb rule is to limit δ to 30^0. With this limit, the maximum power that can be transmitted over a 50 Hz, 300 km long line is the SIL. Thus, for a 400 kV line, it works out to 530 MW. To increase the power flow capability it may be necessary to insert a series capacitor in the line or provide reactive power controller at appropriate locations in the line. The first arrangement is called as series compensation and the second is called shunt (reactive) compensation.

4.4 Voltage and Reactive Power Control

There is an intimate relationship between the voltage in the line and reactive power control. As mentioned earlier, the reactive power requirement is nil while the line supplies SIL. The voltage is also practically constant all along the line (except for small voltage drop due to the line resistance). It is important to avoid transmission of reactive power in the line as this leads to voltage variations and increased losses. If the load at the receiving end draws reactive power (due to motor loads), it is best met by providing reactive power sources adjacent to the loads. As explained in Chapter 2, the reactive power is caused by oscillations in the energy transfer in individual phases. The exchange of energy between the sources and the loads is wasteful as it involves increased current flows in the line without increasing the power flow. However, a transmission line acts as a source (due to the distributed capacitance) and also a load (due to the series inductance) of the reactive power. Assuming that the sending end voltage remains constant and when the receiving end of a line is open-circuited, the excess of reactive power generated by the (distributed) capacitances of the line causes increase in the voltage at the receiving end (this effect is known as Ferranti Effect). For a line of 1000 km length operating at 50 Hz, the receiving end voltage can approach twice the rated value. In addition, the reactive current (reactive power/voltage magnitude) flowing in the line is maximum at the sending end. If generators are connected at the sending end, they have to absorb the reactive power supplied by the line. There is an upper limit to the reactive power absorbed by a synchronous generator dictated by stability considerations and generally this is much less than the upper limit on reactive power flowing out (or supplied) by the generator. The reactive current supplied by the line is called as the charging current. The no-load charging reactive power for a 1000 km long line is, $1.73P_n$ where P_n is the *SIL (Surge Impedance Loading)*.

An effective and simple way of reducing the no-load overvoltages (at fundamental frequency) and line charging currents is to provide shunt reactors. Typically two reactors are connected at the two ends of the line to compensate for about 50% of the charging current. Sometimes, the shunt reactors are connected directly to the line without providing circuit breakers (switches that have to interrupt large currents). Although the cost is reduced by this arrangement, the maximum line loading is reduced. Hence, it is a common practice to switch out shunt reactors when the lines are required to carry higher loadings.

Additional control on the no-load charging current and the overvoltages in the line can be achieved by regulating voltage at one or more locations in the line. If the receiving end voltage is held constant, the no-load charging current for a 1000 km long line is reduced by 1/3 and the total current is shared equally at the two ends of the line (it is assumed that the voltages at the two ends of the line are equal). To regulate voltage at the receiving end, a variable reactive power source is necessary. The reactive power output of this source can swing from negative to positive depending on the power flow in the line. The variable reactive power source can be a bank of shunt connected capacitors and reactors which are mechanically switched, or a synchronous condenser (which is a synchronous motor with no mechanical load). It could also be a Static Var Compensator (SVC) where mechanical switches are replaced by power semiconductor switches for fast control. Fast and continuous control of reactive power to regulate the voltage also increases stability limit substantially. For example, if a transmission line is used to supply a passive load, the maximum power that can be supplied is less than half the maximum power obtained if the load bus voltage is regulated (the actual value of the maximum power depends on the power factor of the load when there is no voltage regulation). It can be as low as 29% of the value with voltage regulation if the power factor is 0.87 (lagging) and increases to a value of 50% if the power factor is unity (note that maximum power flow in a line is independent of the load power factor if the voltages at both ends are regulated). Static compensators using power electronic controllers are members of the FACTS family which is discussed in Chapter 7.

We assume the sending end of a transmission line is connected to the generator (through the step-up transformer) and hence, the voltage at the sending end of the transmission line is regulated. In addition to the receiving end, it is also possible to regulate the voltage at one or more locations inside the line. If the voltages at two ends of a line are regulated, the maximum voltage variation occurs at the midpoint of the line. However, the maximum voltage at the midpoint at no-load will be equal to the maximum voltage at the receiving end of an unregulated line of half the length of the line. If voltage is regulated at midpoint, then the maximum variation occurs at two points, both 1/4 of the line lengths from each end. This clearly shows that we can reduce the effective length of a line by regulating the voltage at one or more locations chosen appropriately. This method of tackling the problems of overvoltage or charging current in generators is called as compensation by line sectioning. A line of 1000 km long supplying a passive load can be made to appear as two 250 km long lines connected in series by regulating the voltages at the receiving end and the midpoint of line. The no-load charging reactive power load on the generator is now reduced to 15.5% of the value corresponding to a line without any compensation. The voltage regulation is provided by any variable reactive power source. The reactive power compensator may have to absorb or supply reactive power depending on the power flow in the line. From considerations of symmetry, (assuming the voltage magnitudes are held constant at the same value at the three locations) it is obvious that the reactive power output of the voltage regulator at the receiving end is same as that the sending end. However, the midpoint voltage regulator has to supply double the reactive power supplied by the voltage regulator at the sending or receiving end.

4.5 Overvoltages in Transmission Lines and Line Design

We have discussed the overvoltages at power frequency in transmission lines due to Ferranti effect. In addition, there are three other types of overvoltages which are of transient nature. These are due to

1. lightning surges
2. switching surges and
3. temporary overvoltages due to load rejection

Overhead lines on tall towers are susceptible to overvoltages caused by lightning flashes. Lightning flashes result when the electric fields between the charged clouds and the earth exceed the breakdown strength of the humid air (less than 15 kV/cm). This results in the movement of negative charges from the cloud toward earth. The path of these charges in the ionized atmosphere is called as the downward leader which progress along an irregular path in steps of about 50 meters. The velocity of the leader is around 100 km/sec. Due to the opposite charges on the earth, an upward leader may arise and when the two leaders meet, a lightning discharge occurs. A visible lightening flash is actually made of several strokes involving movement of charges between the cloud and the ground or sometimes between two clouds. The peak current in a stroke varies from 20,000 A to 2,00,000 A. The median value is about 45,000 A. Although the currents have large peak values, the current rises to the peak value in less than 10 microseconds. After the peak is reached, the current decays exponentially, reaching half the peak value within 20-100 microseconds. When the voltage across the porcelain insulator string exceeds the withstand strength of the air insulation, a flashover occurs between the grounded tower and the conductor resulting in a short circuit of one or more phases. However, if the line is tripped (disconnected from the voltage sources) by opening the circuit breakers at the two ends of the lines, the ionized path (that caused the flashover) regains its insulation strength within one second. If the line is reclosed after some time, it can resume the transfer of power.

Lightning is a major cause of outages (failures) in overhead lines. The overvoltages can be reduced by providing one or two shield wires, also made of ACSR, located above the phase conductors. The objective of providing shield wires is to prevent lightning strokes directly hitting the phase conductors. The overvoltages caused by lightning strokes to the shield wires or the tower can be reduced by reducing the tower footing resistance. To achieve this, ground rods or buried conductors running under the line (called counterpoise) are used.

Switching surges are similar to lightning surges in the wave shape. However, the switching surges rise to the peak value in 250 μs and decay exponentially dropping to half the peak value in 2500 μs. The switching surges are transients caused by closing a circuit breaker (CB) that connects a line (charged or otherwise) to a voltage source. The worst case is when a breaker opens following a short circuit (flashover of the insulator) and then closes within one or two seconds before trapped charges on the line are removed. The peak overvoltages can be as high as 4 times the rated values. They can be reduced by allowing time for the draining of the trapped charges (using grounded reactors) and providing a pre-insertion resistor (within the CB) of suitable value that is connected in series with the line before the breaker contacts close.

It is to be noted that although the lightning and switching surges result in high overvoltages, they are of transient nature (decay fast) and the insulators are able to withstand higher values of these surges compared to the power frequency overvoltages.

The temporary overvoltages caused by load rejection are primarily at power frequency, but may contain harmonics due to the nonlinear magnetization characteristics of the transformers which are left connected to the line where power flow reduces to zero when the circuit breaker at the receiving end trips. These overvoltages are also affected by over speeding of the generator (as the mechanical power input does not become zero suddenly) and also affected by the line capacitances. The problem of ferro-resonance (between the transformer and the shunt capacitances) can complicate the situation. The AVR at the generator can help in reducing the overvoltage if the line charging current is not excessive.

Mechanical Vibrations

The overhead lines are also affected by storms and winds blowing across the line. The phase conductor bundles are suspended by insulator strings at the towers, and the sub-conductors are separated from each other by spacers. The conductors are under tension to minimize the sag of the conductor due to its weight (note that the span length (distance between towers) is about 300 meters). The sub-conductors in a bundle have a diameter ranging from 2.5 cm to 5 cm and the bundle spacing is in the range of 40-50 cm.

There are three types of mechanical vibrations that can arise due to winds. These are

1. Aeolian Vibration
2. Galloping
3. Wake induced oscillation or flutter instability.

The first type of the vibration has frequencies below 20 Hz and amplitude less than 2.5 cm. However, they can cause fatigue of the conductor and the supporting structure. The Aeolian vibrations have to be damped by stock-bridge dampers.

Galloping of conductor is a high amplitude, low frequency phenomenon caused under conditions of icing of conductors in a flat terrain. The frequency does not exceed 1 Hz, but the amplitude may be high enough to cause contact between adjacent phase conductors.

The flutter instability can occur if the bundle spacing is not sufficient and the bundle tilts 5° to 15° with respect to a flat ground surface.

The sag of conductors can increase due to temperature rise and the minimum clearances to ground should be adequate. The growth of vegetation under the line can reduce the required clearances.

4.6 High Voltage AC Cable Transmission

Many people consider overhead transmission as an eyesore and a nuisance. Thus, it would appear that underground cable transmission is quite attractive. Another advantage is that cables are not susceptible to vagaries of weather and thus expected to be more reliable. However, cables are very expensive compared to overhead transmission. Also, a fault in the cable takes lot of time to repair (unlike in the case of overhead lines when most of the faults are self clearing and involves only interruptions of short duration). In addition, AC cables are not suitable for long distance transmission as the charging current becomes excessive.

AC cables are used mainly for

1. to feed distribution substations in metropolitan areas when land is not easily available
2. under water transmission and
3. in switchyards to interconnect transformers and other equipment.

There are four broad categories of cables based on the material used for major insulation.

1. High Pressure Oil Filled (HPOF) paper insulated cables
2. Cables with extruded insulation such as Cross-Linked PolyEthelene (XLPE)
3. Composite laminated insulation having plastic films laminated with Kraft paper, for example PolyPropylene Laminated Paper (PPLP)
4. High pressure Sulphur-Hexafluoride (SF_6) insulated cable designated as GIL

4.7 HVDC Transmission

As mentioned earlier, HVDC transmission has several advantages over AC transmission. These are given below:

1. A two conductor, bipolar (with positive and negative polarities with respect to ground) line can carry approximately same amount of power as a three conductor AC line (for similar current carrying capacity of conductors and insulation levels)
2. A HVDC line does not require reactive power compensation. Also, the power transfer over long distances is unaffected by stability considerations.
3. A bipolar DC line can be operated in the monopolar (with single conductor) mode with ground return. This is not feasible in AC transmission. A three phase transmission cannot operate with disconnection of one or two phases (due to high ground impedances).
4. A DC line can interconnect two AC systems operating at two different frequencies.
5. Since DC line does not require charging currents to operate, DC cables, particularly under water, are superior over AC cables and have no limitations on the distance.
6. The power electronic converters at the two ends of a DC line are fast acting and permit rapid control of power flow. This capability can be utilized to improve the system security and stability

The implications of the first advantage is that DC line is more economical than AC. It requires less Right of Way (RoW), reduced costs of construction and reduced power losses. However, the reduced costs have to be balanced against the additional costs of conversion (and the losses associated with it) at two ends of the line. There is a break-even distance above which DC line is more economical than AC line. It is in the range of 500-800 km.

The implication of the second advantage is that a DC line can carry more power than an AC line for long distance transmission. Thus, the power transfer capability of a ±500 kV DC line has gone up from 1500 MW to 3000 MW as the thyristor device ratings have increased over the years. The advantage number 6 implies that the margins required for a DC line are minimal and depend on the power modulation capability that is designed for the DC line with the intention of improving the AC system performance.

Advantage number 3 implies that a DC line can be more reliable than a single circuit AC line. If a fault results in the loss of one pole of a bipolar DC line, the remaining pole can transfer partial or full power depending on the DC line design. Advantage number 4 implies that DC is the only feasible solution in interconnecting two systems of different frequencies (Japan has both 50 and 60Hz systems in different islands forming the country). Even when the nominal frequency is same in two regions, the actual operating frequencies may differ if load frequency control(LFC) is absent or not practiced. The asynchronous DC interconnections are normally of the Back-To-Back (BTB) type where both converters (conversion from AC to DC and from DC to AC) are located in the same building with no DC line. Sometimes, such stations are called as frequency converters. In India, the rotary frequency converter at the oldest hydropower station at Shiva Samudram was replaced by BTB DC link.

Advantage number 5 implies that whenever sea crossing is involved, HVDC cable is the only feasible solution. A major example is the 2000 MW cross channel link connecting England and France. This link transfers the surplus nuclear power from France to England.

HVDC also has few disadvantages listed below:

1. The cost of conversion
2. Generation of AC current and DC voltage harmonics which require filtering
3. Problems of commutation failures in thyristor converters which tend to be more severe when a HVDC link is connected to weak AC grids
4. Converter stations require reactive power supply that increase with the power flow in the link
5. DC breakers are costly and voltage transformation is not feasible.

Because of these disadvantages, HVDC lines are used mainly for point to point transmission of power, whenever the advantages score over AC. The advent of power trading has increased the demand for HVDC option as the power flow in the line can be regulated independent of the system conditions.

The emerging high power electronic controllers using Voltage Source Converters (VSC) have some advantages over the Current Source Converters (CSC) using thyristor devices. The VSC converters use IGBT devices and have been applied for HVDC transmission for modest ratings (up to 400 MW). The major applications are

1. power supply to urban areas using underground cables
2. power transfer from off-shore wind farms and
3. supplying power to off-shore platforms for drilling oil

In the recent times (after 2010) by applying Modular Multilevel Converter (MMC) based VSC-HVDC transmission, several links with capacities up to 1000 MW (at +/- 500 kV) are under construction to evacuate offshore wind power.

4.8 Electro-Magnetic Fields (EMF) Associated with Transmission Lines

In recent years there is concern about EMF at very high frequencies and their effects on health. There has also been concern about Extra Low Frequency (ELF) EMF caused by EHV lines. The major issues are the safety issues as the RoW is not fenced and hence vehicles (including trucks) may be parked under a line or animals may graze. The electric fields at the ground level may exceed 10 kV/meter for EHV lines. It was established from the Russian experience that there could be long term effects on health if the AC electric fields exceed 5 kV/meter.

In general, there are 4 types of effects for human beings. These are:

Direct Contact Currents: These are currents flowing through the human body when a person contacts an object which has an induced voltage as a result of the electric field.

Spark Discharges: These occur when one or more objects come into close proximity with a metallic object which is charged.

Direct Effects: An electric field may directly affect a person by stimulation of body hair and induce circulating currents that may penetrate below the skin surface.

Magnetic Fields: Can cause induced currents in a person's body due to electromagnetic coupling with the line.

The induced currents are primarily due to the time variation of the electric fields and hence it may appear that HVDC lines do not have similar effects as AC lines. However, DC lines also have a space charge field caused by ionization, in addition to the static fields.

The basic ionization process is same for AC and DC conductors, except that, in the AC case, nearly all the ions created during one half cycle are recombined when the polarity reverses in the next half cycle. Very few ions escape the ionization zone near the conductor at power frequencies. However, with DC conductors some of the ions migrate toward earth. As a result, the electric field near the ground is increased. For DC lines, the 'induced current' on a person under the line is the ion current interrupted by that person.

The effects if AC and DC currents on human beings, based on measurements is given in Table 4.2.

Table 4.2 Physiological effects and thresholds for body currents and shock currents

	Direct Current (mA)		60 Hz (RMS) (mA)	
	Men	**Women**	**Men**	**Women**
1. 'Threshold of Perception' level	5.2	3.5	1.1	0.7
2. Uncomfortable shock, but not painful	9.0	6.0	1.8	1.2
3. Painful shock ;safe let-go' level	62	41	9.0	6.0

(Data taken form BPA 'Transmission Line References Book - HVDC to ±600 kV EPRI 1977)

Appendix B gives a brief note on electric shocks and safety issues.

According to World Health Organization (WHO), static magnetic fields, measured in terms of flux density due to the natural geomagnetic field over the earth's surface, varies between 0.035-0.07 millli Tesla (350-700 milli Gauss). Recent technological innovations used in research and medical applications (for scanning) have higher magnetic fields, up to 10 T. For static magnetic fields, acute effects are only likely to occur when there is movement such as motion of a person or internal body movement, such as blood flow or heart beat. Moving within a field above 2 T can experience vertigo and nausea. It is not established whether static magnetic fields in the milliTesla range have long-term health effects.

The effects of static electric fields in the long term have not been properly investigated.

4.9 Distribution of Power

Distribution lines (also called as feeders) are the final link in the delivery of power from generating stations to the retail customer (residential, commercial or industrial). The HV or EHV transmission lines carry power to a distribution substation where a step down transformer converts the voltage to medium voltages (2.4 -34.5 kV). From the feeders, further stepping down takes place at or near the customer premises to low voltage distribution (also called as secondary) from 120 to 600 V. In Europe and India, the basic utilization voltage for residential and commercial loads is 400 V (phase to phase) and 230 V (phase to neutral) in three phase supply. The distribution transformer secondary has wye connection with grounded neutral. To reduce installation costs, small distribution transformers are pole mounted in overhead distribution. In underground distribution, transformers are installed in street vaults, in manholes direct-buried, on pads at ground level or within buildings. The service or secondary conductors are mostly aluminum conductors (instead of copper) due to cost considerations.

Just as SCADA system is used for supervisory control and data acquisition in bulk power systems (combined generation and transmission systems), distribution automation and control (DAC) system is used for distribution systems. Its functions include load switching, load shedding, load reconfiguration to enable routine maintenance on feeders, voltage regulation, transformer and feeder load management, capacitor control, fault detection and isolation, implementing automatic meter reading and time-of-day pricing.

The power losses in a distribution system are primarily due to the resistance of the conductors and no load losses in the transformers that result in the temperature rise that should be regulated to avoid damage to insulation. While the power losses in a AC transmission system are typically less than 5 %, the distribution power losses can be contained by suitable design to a value below 10 %. It is to be noted that the losses refer to the 'technical' losses and not 'commercial' losses which are mainly due to faulty energy meters and/or power thefts.

CHAPTER 5

Power System Operation and Basic Controls

"The whole is greater than the sum of its parts"
Max Wertheimer

5.1 Introduction

In this chapter, we will discuss the system operation under normal conditions and the basic control functions that enable the supply of electricity to the customers at different locations in the system. In many countries, different utilities also operate as pools within which electric power is traded among the members.

As mentioned in Chapter 1, the electric energy is not stored and it is required to match the generation of power to the load demand that varies with the time of the day. A typical daily load curve is shown in Fig. 5.1.

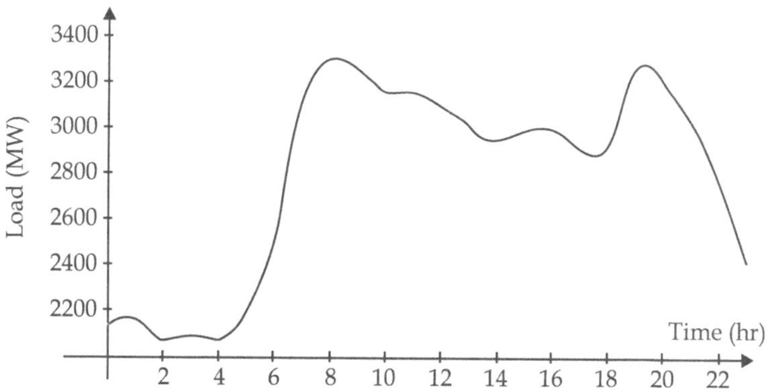

Fig. 5.1 A typical daily load curve

The daily load variation is cyclic and repeats on a daily basis although the peak load may change from day to day. For example, during weekends and holidays the peak load will be less. There could also be seasonal variations depending on the temperature. However, the shape of a daily load curve does not vary too much. There is a minimum load (called as base load) which is met by base-loaded generators operating at their rated power for 24 hours. Nuclear plants and large coal-fired units are normally base-loaded due to their operating characteristics which do not permit fast change in their power outputs or frequent shut down. The variable part of the load is carried by plants that can be controlled. Typically medium sized fossil fuel fired units and hydro generators are used for control of generation to meet the load requirements. During peak load, smaller and costlier plants (operating with gas turbines and diesel generators) are operated which are easier to start and shut down each day. In addition, some generators that can be controlled operate at partial capacity to provide 'spinning reserve'. Since the variation of load cannot be predicted exactly, it is necessary to have sufficient spinning reserve that will be able to supply a sudden increase in the power demand or to meet the contingency of a sudden tripping of a generator due to forced outage or fault. We can pose the following questions in this context.

1. How does the system operator decides the quantum of the total generation required that should be equal to the total load plus power losses in the transmission network? We assume here, the load demand is measured at the receiving end substations connected to the transmission network and hence will include the losses in the distribution system.
2. How is the required total generation distributed among the various generators in the system?
3. How are transmission bottlenecks (or congestion) handled?
4. How does reactive power balance achieved?

With regard to the last question, it needs to be mentioned that load (power demand) has two components - (i) active power (P) that supplies the electric energy and (ii) reactive power (Q) that does not supply electric energy, but is required to maintain magnetic fields in electric motors and other devices (also required to maintain magnetic fields surrounding the conductors of distribution circuits). Shunt connected capacitors connected at the distribution stations can be used to supply the reactive power demands associated with loads. It is not desirable to supply reactive power over long distances as the receiving end voltages are reduced which reduce the maximum power that can be supplied and also increase power losses.

We will discuss the answers to the 3rd question little later. The answer to the 2nd question is that the outputs of individual generators are primarily determined by the economic factors. We want to minimize the total cost of production of power at any given time. The cost of production of power by a single generator is mainly determined by the fuel costs that are influenced by the efficiency of power generation. For thermal power stations using fossil fuels (coal, oil and gas), the efficiency is expressed as heat rate (BTU/kilo Watt Hour). BTU stands for British Thermal Unit and is equal to 1054 Joules. The heat rate varies from about 9,000 to 10,000 BTU/kWh. The efficiency is about 35% in converting the chemical energy. Combined cycle plants that use high temperature exhaust gases from one or more gas turbines to generate steam in heat-recovery steam generators have efficiencies over 50%.

The efficiency of thermal generation is not a constant, but a function of the power output. If the efficiencies are constant and the fuel used is identical within a region (where differences in the transportation costs of the fuel can be ignored), the generators should be loaded according to a merit order - the generator with the highest efficiency is loaded fully before other generators are loaded according to their merit order based on the efficiencies. However, in thermal generation there are constraints posed by the burning of the fuel that require a minimum output for a generator that depends upon the stability of fuel combustion and steam generator design. Thus, in a supercritical boiler, the minimum power output could be 30% of the rated power. When several generators are supplying power to meet the required load, the power outputs of the generators are determined by their incremental heat rates or costs of production.

In comparison with thermal generation, hydroelectric generation does not involve any fuel consumption. However, the quantum of water available for generation may be limited when the inflow into the reservoir is constrained. In rain-fed reservoirs, the water level in the reservoir varies depending on the monsoon rains. The reservoirs may be coupled hydraulically when the water flowing out of one plant flows into the reservoirs connected with downstream plants.

The potential (mechanical) energy stored in the water is converted into the electrical energy in a hydroelectric plant where a hydraulic turbine drives the low speed (salient pole) generators.

A flow of one cubic feet of water per second, falling 100 feet can generate about 8 kW assuming the efficiency at 95%. The efficiency is not a constant, but varies with the load. The best efficiency is obtained for a particular loading at the designed head. Maximum efficiency varies between 90-95%.

The dispatch of hydroelectric generation is also dependent on the use of the reservoir for irrigation in addition to power generation. There could be other constraints involving river transportation and fishing. Hence, in dispatching the hydro plants, the objective is to maximize the energy generated over the yearly water cycle. The operating costs are not very relevant. Incidentally, some hydro stations (pumped-storage hydro) can function as energy storage by operating as motors pumping water to the reservoir at a higher elevation during periods of low loads (in the nights, for example). They work as generators during the peak loads thereby reducing the requirements of the total installed capacity.

5.2 Control of Generation

Although it would appear to be a complex task to ensure that the total generation match the total load in the system plus the losses, the task is made easier by the two facts –

1. The control of power flow in a AC line does not require an external controller – it is the natural characteristic of a AC line
2. The AC system containing synchronous generators has to operate in synchronism. The average system frequency is same throughout the system.

The control of generation in a power system can be explained by first considering the example of a single power station supplying power to a load centre (represented as common bus where all the loads are connected in parallel (see Fig. 5.2).

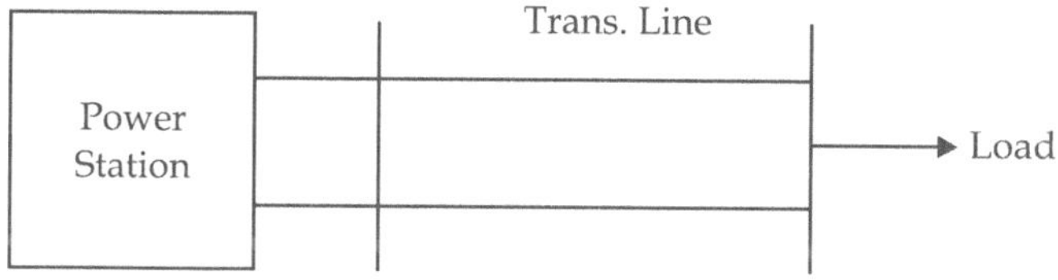

Fig. 5.2 A power station supplying power to a load centre

Note that there can be two or more lines transferring power from the power station to the load. From considerations of reliability, the loss of any one transmission line (due to fault) should not result in overloading of the remaining lines in parallel.

Any increase in the power demand at the load centre will be met automatically by the transmission network by increasing the angular difference between the sending and the receiving end. This does not require any external controller. The increase in the electric power output of the generator (assuming a single generator) will cause a drop in the speed of the generator rotor driven by the turbine (steam or hydro). To ensure that the speed (and hence the electrical frequency) is brought back to its original value, a speed governor is provided which senses the speed and controls the position of the control valves (for a steam turbine) or gate (for a hydro turbine). Since there are multiple turbines in a thermal power station, the control valves are located before the steam enters the higher pressure turbine. For large increase in the load, it would be essential to regulate the boiler pressure by controlling the fuel ignition system. The speed governor is a mechanical or electronic device for speed sensing and electro-hydraulic motors for controlling the valve/gate position. By providing an integrator in the control loop, isochronous operation of the governor is obtained with frequency regulation. This is feasible only when a single generator feeds the load.

Typically several generators are connected in parallel in one power station (if more than one power station is considered we can still view the generators operating in parallel in the sense that they are subjected to same deviation in the average frequency). To ensure that the parallel connected generators share the load in predetermined ratios, the speed governor control is given a drooping characteristic such that the speed drops as the power output increases.

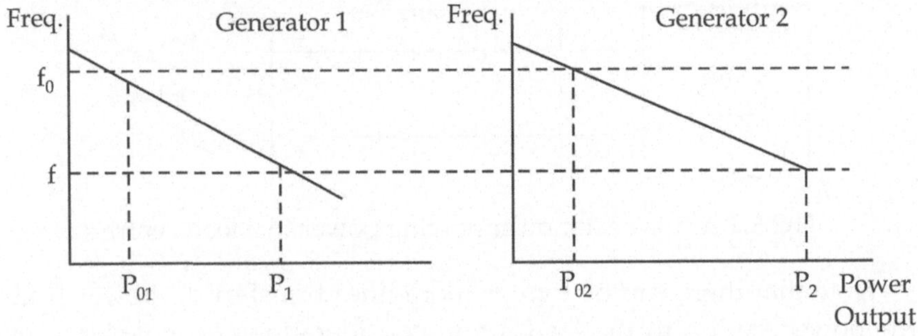

Fig. 5.3 Speed governor characteristics

Fig 5.3 shows the variation of the frequency as a linear function of power output for two generators (1 and 2). f_0 is the nominal frequency and f is the operating frequency when power output is increased. The governor for generator 1 has a higher speed droop than that of generator 2. The generator 2 takes a higher share of the increase in the load as compared to generator 1.

Typically the power output is expressed in per unit (or percentage) of the rated power of the generator. If the speed droop (expressed in percentage change in the speed when the power changes by 100% of the rating of a generator) is identical for all the generators in a power station, they share the load in a manner such that any increase in the load is met by a generator, proportional to its rating. The speed droop varies in the range of 3-5%. A 5% speed droop implies that the frequency varies by 5% as the power increases from zero to 100% of the rating.

We can vary the output of a generator without affecting the frequency by changing the setting (reference speed or frequency) of the speed changer. Fig. 5.4 shows how the power output of a generator can be increased without affecting the operating frequency. What it implies is that it is possible to load the generators independent of the governor speed droop characteristics. However, this does not mean the governor action is not important. Actually, immediately after a load-generation imbalance, the governor action (primary control) comes into play to stabilize the frequency. The speed changer setting is controlled by a higher level (secondary) power-frequency or load frequency control (LFC).

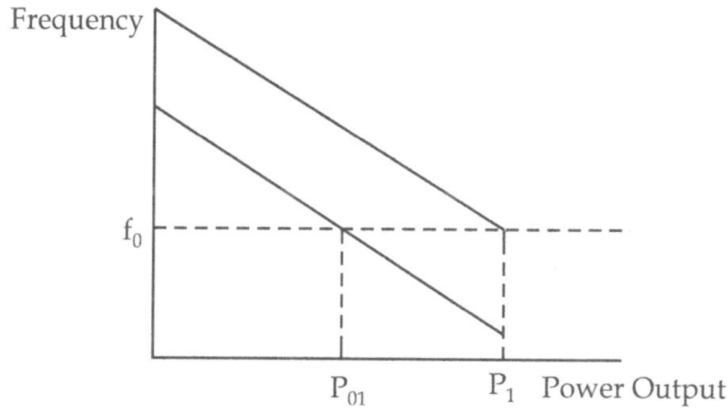

Fig. 5.4 Power frequency characteristics

It is to be noted that in a large system consisting of hundreds of generators, increase in load can be met by frequency regulation at selected generators which have spinning reserve and can raise generation at sufficiently high rates (such as hydro generators). Where hydro generators are not adequate, thermal generators are also utilized with rate constraints.

5.3 Effects of Load Characteristics

The load characteristics of motor loads are such that the power demand is reduced when the frequency is reduced. This assists the governor action and helps in reducing the frequency drop when a sudden increase in the load occurs.

If the generation exceeds the load, the frequency increases which is limited only by the governor action and the load characteristics. The secondary control (or regulation) of frequency using Load Frequency Control (LFC) is practically absent in several utilities in developing countries including India. If governor action is also blocked, this results in unreasonable increase in frequency at times of minimum load that can affect customer equipment such as adjustable speed drives. If generation is not reduced during the periods of low loads, (during the daily load cycle) because of perverse incentives to maintain high plant load factor, the situation can further aggravate. In eastern region of India frequencies rising to 53 Hz was a common feature until the Availability Based Tariff (ABT) was introduced. ABT scheme is discussed in chapter 9.

Automatic Generation Control (AGC)

With the introduction of modern energy control centres following the major Northeast blackout in USA in November 1965, the automatic generation control using direct digital control action based on measurements of generator power outputs and tie-line power flow, has been introduced.

AGC has three major objectives

1. To regulate the system frequency at nominal volume (50 or 60 Hz).
2. To maintain the schedules of interchange of power between control areas.
3. To ensure economic dispatch that will minimize production costs.

The AGC is applied to a part of the overall system, designated as the control area in which the AGC is applied. The selection of control area may be based on the ownership or other considerations when RTO (Regional Transmission Operator) or ISO (Independent System Operator) function in a restructured power system.

The objectives 1 and 3 have been discussed earlier. Let us now consider the second objective. A control area is normally connected to adjacent control areas by interconnecting (tie) lines. A control area may be importing or exporting power to the other areas as a firm (contractual basis) or wheeling power from one area to another (again based on prior agreements). In both cases, the net interchange of power (sum of power exported over the tie lines) must be kept constant under normal conditions (note: this assumes the power losses in the system are approximately constant).

Now, let us consider what happens when there is a sudden increase in the load of control area 1 (see Fig 5.5). Fig 5.5 shows a two area system with a tie line.

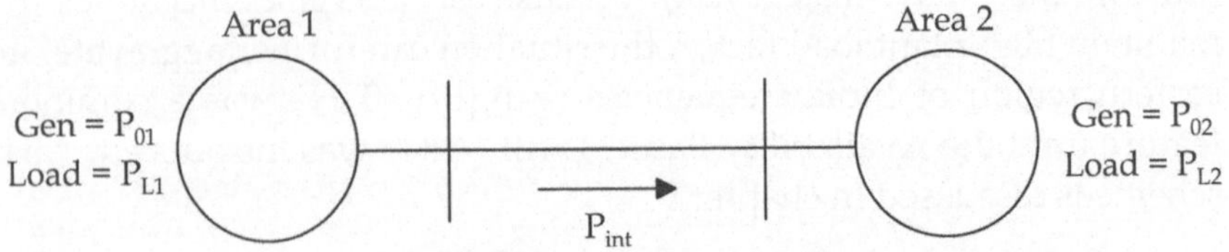

Fig. 5.5 A two area system

Until the load balance is restored by increasing generation in area 1, the drop in the frequency results in the reduction of power interchange (P_{int}) which is the power flowing over the tie line from area 1 to area 2 (note this is an automatic action in a AC grid due to the inherent characteristics of AC lines).

The increase in the generation in control area 1 is determined such that (a) frequency is regulated and (b) power interchanges are retained at the predetermined values. The AGC action must be limited to only the area where a disturbance has occurred. Thus for load increase in area 1, the AGC in area 2 should not act (unnecessarily).

A simple control law (applicable to all the control areas) determines the Area Control Errors (ACE) as follows:

$$ACE = -(\Delta P_{int} + B\Delta f)$$

Where ΔP_{int} is the net increase in the power flow (flowing out of the control area) and Δf is the increase in the average frequency. B is called 'Frequency Bias Factor'. The ACE is equal to the increase in the required generation.

The average frequency is same at all buses in an interconnected power system. Thus, measurements of both frequency and the power interchange are local to a particular area (no measurements from other areas is required).

For a two area system shown in Fig. 5.5, there are four possibilities as given below:

Case 1: $\Delta P_{L1} > 0, \Delta f < 0, \Delta P_{int} < 0, \Delta P_{L2} = 0$

Action: Increase generation in area 1

Case 2: $\Delta P_{L1} < 0, \Delta f > 0, \Delta P_{int} > 0, \Delta P_{L2} = 0$

Action: Reduce generation in area 1

Case 3: $\Delta P_{L2} > 0, \Delta f < 0, \Delta P_{int} > 0, \Delta P_{L1} = 0$

Action: Increase generation in area 2

Case 4: $\Delta P_{L2} < 0, \Delta f > 0, \Delta P_{int} < 0, \Delta P_{L1} = 0$

Action: Reduce generation in area 2.

Note that ΔP_{L1} and ΔP_{L2} are increments in the load (powers) in areas 1 and 2 respectively.

The frequency bias factor (B) for each area has to be chosen by the consideration of the generating units operating at any given time and the speed droop characteristics of the speed governors associated with the operating units. While the area controller (ACE) determines the total generation raise in a specified control area, the allocation of power generation to individual unit is determined by the participation factors based on economic dispatch .The reference signal (in terms of desired power) for the speed changer setting is determined from the relation

$$P_{refi} = P_{basei} + F_i \Delta P_t$$

where P_{refi}, P_{basei} and F_i are the reference value, base value and participation factor for unit *i*. ΔP_t is the total change in the generation required. The power losses in the system are accounted in computing the total change in the generation. P_{basei} is modified every few minutes from economic dispatch. The ACE is calculated at intervals of few seconds (5-10 s).

In implementing AGC, following points are also considered:

1. When ACE is large, more units are controlled initially to overcome the limitations of rate constraints (in increasing generation). Once the ACE is reduced to zero, the AGC acts to restore economic dispatch.
2. The ACE signal is filtered to avoid random fluctuations in generation which are small in magnitude.
3. AGC must take into account failures in telemetry and also failures in the control action at the generators and take appropriate steps to avoid errors. In case of telemetry failure, the AGC action is suspended. If the power controller at a generating unit is not responding, the control action is reallocated to another unit.
4. AGC must take into account the base-loaded units that do not participate in the AGC and allow manual control of some units.

5.4 Demand Side Management (DSM) and Load Shedding for Frequency Regulation

If the generation is not adequate to supply the demand, it is possible to apply Demand Side Management (DSM), also called as Demand

Response (DR) in the recent smart grid literature, by the system operator with prior arrangements with the large customers who have heating/cooling loads which can be switched off for some time. The switching off and on operations are done by the system operator by sending a signal under what was called as ripple control system. The heating or cooling loads can withstand temporary shut downs as the temperature does not change fast and when deviations in the temperature settings are tolerable. If customers have diesel generating units, they can be started when there is power failure and the switching on/off of these units can also be controlled by the system operator. Generally these operations will occur at low or medium voltage distribution systems and can be viewed as part of the distribution automation or smart grid. Instead of diesel generating units, renewable generation such as solar photovoltaic or wind power generation in conjunction with battery or fuel cell storage systems can be used to supplement power supplied by the main grid. If the total generation (including the distributed generation at the customer premises) is inadequate to regulate the frequency, then load shedding is necessary. Only loads that are not considered important are shed in response to the frequency signal. The load shedding is carried out as a last resort and the amounts of load to be shed as the frequency drops, have to be carefully worked out to minimize the interruptions. It is obvious that the load shedding is to be carried out only in emergency type situations. In contrast, DSM is based on the contract between the supplier and the customer. The price of energy for load under DSM will be less than what is normally charged.

5.5 Control of Reactive Power and Voltage

Apart from the control of power and frequency in a system, the other major control function is to control reactive power and voltage at different locations in the system. The generator terminal voltage is regulated by using field excitation controller. The fast control of field excitation requires field forcing as the changes in the flux linking the field winding involve few seconds to take effect (due to large inductance of the field winding). The field current required to establish rated terminal voltage on no-load in the generator is about half the value required to circulate rated armature current (in the stator windings) when the stator windings are short-circuited.

Field forcing means that a large field voltage is initially applied to force rapid increase in the field current. The AVR (Automatic Voltage Regulator which is an electronic amplifier with a high gain) and a fast-acting static exciter using power electronic converter that is fed from a transformer connected to the generator terminals, ensure a high response excitation control to improve the performance of large modern turbo generators. As a matter of fact, without the high response excitation control system, the generator will not be able to supply the rated power even if the load is directly connected at the generator terminals. With large generators of 1000 MVA rating, the power is usually transmitted by EHV transmission lines to large load centres located at distances of hundreds of kilometers. Maintaining stability is of prime importance in such cases. Stability issues will be discussed in the next chapter.

Voltage profile in a AC grid, is determined by the adjustments in the terminal voltage of generators, on-load tap changers of the transformers in the generating and receiving end substations, in addition to the injection of reactive power at the specific locations in the high voltage grid. The reactive power sources are -mechanically switched capacitors, synchronous condensers and static var compensators (SVC) that use power electronic switches for fast control of reactive power. Under normal conditions, there is an optimum choice of the operating values of the scattered reactive sources and voltage settings of the automatic voltage regulators that will help to minimize the losses in the AC transmission grid. Even a saving of one percent in the power losses can bring in substantial savings in the fuel costs and justify the use of optimum power flow algorithm that includes both economic dispatch (selection of generator power outputs) and minimization of power losses. Actually both objectives can be clubbed together. It is also possible to consider the constraints in the system operation. Firstly, the control variables have an operating range that defines the minimum and maximum limits. Secondly, it is necessary to ensure that any equipment in the system, transmission lines, transformers and the generators are not overloaded. The upper limits on the reactive power outputs of generators are based on the limits on the field excitation (determined by the heating of the field coils or windings).

5.6 Operational Planning

This is different from long term or medium term planning whose horizon may extend up to 10 years. For the operational planning, it is typically about one week and involves following activities:

1. Short term load forecasting.
2. Unit commitment (deciding the selection of generating units that have to be operational, taking into account both forced and maintenance outages of existing units).
3. Hydrothermal coordination to decide the optimum combination of hydro and thermal generation. This also includes planning of economic dispatch.

5.7 Energy Control Centre

After the 1965 Northeast blackout in USA, energy control centres introduced Supervisory Control And Data Acquisition (SCADA) in a big way to monitor the power flows and voltage magnitudes in critical lines and buses. The data on circuit breaks status (on or off) is also necessary to check the network topology.

If the voltage magnitudes and phase angles are known at all buses in the system, accurately, it is possible to compute the power and reactive power flows in the entire system. There are (2N-1) variables of voltage magnitudes and angles in a system with 'N' buses (note that one angle at an arbitrarily chosen bus can be assumed as zero). These variables are called as state variables and their estimation is performed on-line, utilizing measurements. Statistical techniques are used for the 'state estimation' which enables online load (or power) flow analysis to be carried out to determine the power flows in all the lines and the power injections at all buses. This computation is necessary to check whether any line or transformer is overloaded. This is called as security monitoring.

A power system is said to be operating in a 'Normal State' if it is able to supply the load demand (both power and reactive power) without causing overloading any equipment or transmission lines. Also, the voltages at every bus should be satisfactory (lie in a range of 0.95 to 1.05 times the rated voltage).

If there is an overload or unsatisfactory voltage, the system is said to be in 'emergency state'. If no action is taken in a reasonable time, protection system will operate to trip the equipment or line which may result in cascading outages and break-up of the system.

There are two types of crises that result in transition of a system from the 'Normal' to 'Emergency' state. The first one is called 'viability crisis' described in the previous paragraph. The second crisis is called 'stability crisis' which results in loss of synchronism among generators or voltage collapse at some load buses. The control action to prevent transition to emergency state is called as the preventive control. On the detection of an emergency, the control action to force the system to a normal state is called as emergency control. This will be discussed in the next chapter.

Often, the time required to initiate emergency control may not be adequate to prevent the collapse of the system and wide-spread blackout. Hence preventive control is used in systems when generation facilities are sufficient. However, inadequate transmission facilities can limit the effectiveness of the preventive control action.

The application of the preventive control involves first asking question – 'what if a contingency (of tripping of a generator or a transmission line) occurs and what are its effects' .If the effects involve transition to an emergency state, we say that the system operation (even under normal conditions) violates "security" constraints. Generally, only single contingencies are considered and are ranked according to their severity. The prediction of viability crisis is based on performing power flow analysis. The prediction of stability crisis requires stability analysis.

If a contingency results in a viability crisis, the preventive control is aimed at rescheduling of generation to avoid the crisis (note that the crisis has not occurred as the contingency is only imagined. However, if the contingency occurs, preventive control will ensure that the crisis will be avoided). The major problem in this approach is that its is impossible to consider all possible contingencies in advance. Also, the prediction may not be accurate and hence the results may not be satisfactory if the contingency occurs. Note also the fact that the preventive control can increase production costs as the rescheduling of the generation results in departure from economic dispatch.

The contingency analysis is a useful tool even if preventive control is not adopted. Firstly, it helps to assess the risks in the operation of power systems. Secondly, the system operator can be prepared to handle a contingency if he/she knows the effects in advance.

CHAPTER 6

System Protection and Collapse

"Chaos is the score upon which reality is written"

Henry Miller

6.1 Introduction

The electrical energy in power systems is produced just in time and is also utilized practically instantaneously. This implies that any breakdown in the system, whether in generating stations, substations, transmission lines or distribution feeders has to be attended immediately and rectified. If there is a fault in a transmission line, it has to be detected very fast and the line should be disconnected using fast acting circuit breakers (CB) at the two ends of the line. Most of the faults are caused by short circuits due to flash over of the air insulation between the conductor and the ground or between any two conductors. In overhead transmission lines, the insulation will recover within a second or more and the line can be reinserted by closing the CB. If the fault persists, even after three attempts to clear the fault, the line is kept out of service until the fault (which may be permanent due to, say, reduced clearances caused by increased sag of the conductor due to higher temperature or any other reason) is rectified. The detection of fault is done by protective relaying which can act within one cycle (20 ms in a 50 Hz system) and the breaker can interrupt fault currents within 2 cycles (40 ms). The circuit breakers are similar to the switches, one uses in homes. However, these switches carry only a few amperes and operate at 110 or 220 V. We do not even notice, that while opening a switch carrying current, there is a tiny arc due to the inductance in the circuit which creates a transient overvoltage. Unlike in homes, the switches (telephone relays) used in telecommunication circuits operate at frequencies up to 100 operations per second and have long life (up to a billion operations) but have to handle only very low currents and voltages (milliamperes and tens of volts). They are carefully protected in clean rooms. Control switches (for motors and other electric equipment) have to carry heavier currents (up to kiloamperes), but still at low voltages (hundreds of volts). The switching frequencies are typically few operations per minute and the switches last for tens of millions of operations.

Circuit breakers in transmission networks may carry continuous currents of few thousand amperes at service voltages up to 800 kV. Short circuit currents of 70,000 to 90,000 A must be interrupted reliably within 2 to 3 cycles (33 to 50 ms in 60 Hz system). The generator circuit breakers may carry as much as 36,000 A continuously and may be subjected to short circuit currents up to150,000 A. For outdoor breakers, the ambient conditions can be very severe and yet the maintenance work must be minimum.

6.2 High Voltage Circuit Breakers

Regardless of the application, any breaker or switch must accomplish the following duties:

1. When closed, it should be an ideal conductor
2. When open, it should be an ideal insulator
3. When closed, it should be able to interrupt its assigned current promptly at any instant without causing dangerous over voltages. In some cases it must limit a fault current to tolerable value.
4. When open, it should be able to close promptly at any instant without any problems like contact welding.

Switching Arc

In the switches at our homes, we have to merely open the contacts a fraction of a millimeter. A small arc causes the current to continue until the next zero of the AC current. Within about a microsecond of the current zero, a space charge layer able to withstand about 200 V is built up at the new cathodes (from where the electrons leave). For higher voltages, either more gaps in series must be used or the arc has to be lengthened until the voltage across it exceeds the value required for its survival. In this way, the current is continuously reduced to zero by increasing the arc resistance.

For voltages of the order of kilovolts, the arc lengths required for interruption would be excessive (about 1 meter per kV). This may indicate that switching arc is a major problem in interruption. But it must be understood, without the arc, the voltage at the moment of contact separation (proportional to the derivative of the current in an inductive circuit) will be excessive. Actually, the arc preserves the current until the energy $\left(\frac{1}{2}Li^2\right)$ stored in the inductance L, disappears by itself without any damage, at the next current zero.

To avoid arcing, the contacts must be opened exactly at current zero. But it is extremely difficult to predict the moment with sufficient accuracy and speed. To avoid an arc, the current must be less than 0.3A. A sinusoidal current (50 Hz frequency) of 100 kA rms reaches this value at 10 nanoseconds before its zero. Thereafter the restriking voltage arises at the rate of about 10 kV per microsecond. To avoid the restriking of the arc, a 10^{-5} *cm* contact gap is required after 1 ns. This implies an acceleration of 10^{10} g (where g stands for acceleration due to gravity). As a matter of fact, the problem of synchronization is solved reliably by the arc itself because it ceases spontaneously at the next current zero of the current; provided re-ignition caused by the thermal inertia is avoided.

Since synchronous switching by mechanical means is not feasible, the question can be posed whether the arc is the only alternative and if so why? The arc plasma is the only substance able to change its conductivity from 10 mho/cm (a good conductor) to 10^{-12} mho/cm (a good insulator) merely when the temperature changes from 10,000 to 1,000 degree Kelvin. Moreover, the change from the conducting to isolating state occurs extremely rapidly. In contrast, the conductances of semiconductor diodes and thyristors vary only by 8 orders and no faster than that by an arc. They are neither true conductors nor insulators and are extremely sensitive to current or voltage overload. A plasma (forming the arc) has never been destroyed by overload.

The crucial problems of the switching arc are

1. The mechanism of change of conductance.

2. Means of reducing its time constant.
3. The avoidance of re-ignition after current zero.

The conductivity of the arc column is affected by its temperature only, which is about 6000° *K* in open-air arcs of a few amperes, but probably up to 25000° *K* and more in high power circuit breakers at the current maximum. The conductivity is affected by ionization of the arc medium. The thermal ionization resulting from random collisions in a hot gas, must be distinguished from impact ionization, which is caused by electrons accelerated in an electric field between two collisions. The latter cause dielectric breakdown, even in a cold gas. The dielectric breakdown is affected by the dielectric strength. Air as an insulator breaks down at 30 kV/cm at normal pressure. Compressed air and oil have much higher dielectric strength than air at normal pressure.

The arc medium used in circuit breakers is oil or gas under pressure. Air and sulphur hexafluoride (SF_6) gas are used. Vacuum breakers are also used in medium voltage applications.

6.3 Effects of Lightning Strikes

Lightning bolt appears when the voltage between the cloud and the earth becomes high enough to release electrons from the atoms in the air and create electric spark. This will often hit the steel tower of a transmission line because the top of the tower is the highest conductor in the vicinity. The high current in the lightning will raise the voltage of the steel tower such that an electric spark will appear across one of the strings of the porcelain insulators which are hung from the steel tower and which support the conductors. This spark will create a conducting path through the air from the supporting metal on the steel tower (at the top of the string), to the transmission line conductor. This electric spark immediately turns into an arc because of power flow from the conductor to the ground (through the tower). The circuit breaker protecting the line will remove the source of power long enough for the arc to cool down.

Another way in which lightning can cause creation of an arc is for the lightning to strike a transmission line conductor. The charges from the lightning stroke will travel down the transmission line at speed approaching that of light. The travelling wave, when it hits the terminals of the line, gets reflected. The resultant voltage may be double the line voltage under certain conditions. This high voltage may cause an electric spark or internal breakdown of the insulation of a transformer or circuit breaker. The lightning arresters can help to prevent such occurrences.

In the absence of fast acting circuit breakers, lightning induced arcs can cause porcelain insulators to crack or damage the surface. Metal is vaporized so violently from the ends of the arc that nearby insulation and other surfaces are often covered with molten metal.

If the flow of power to an arc is removed, the air will cool quite rapidly (within a quarter of a cycle).

6.4 Protection against Transient Overvoltages

The transient over voltages are classified into three categories as follows:

1. Impulse voltage which rise to the peak within 0.3 to 3 microseconds and decay within 100 microseconds.
2. Switching overvoltages which rise to the peak in more than 100 microseconds and decay within 3 milliseconds.
3. Temporary overvoltages that last few seconds.

The impulse voltages are usually caused by lightning. The switching overvoltages as the name suggests, are caused by switching action. In AC transmission line, the most onerous condition arises while reclosing of the breakers after clearing of the fault and when there are still trapped charges on the line due to line capacitances. Installing a (pre-insertion) resistor that is first connected across the breaker contacts before the breaker is closed, can reduce these overvoltages.

The temporary over voltages are of supply frequency and are usually caused by load rejection on a line. The so called Ferranti effect (due to line capacitances), the non-linearity of the magnetic circuit in transformers and the phenomenon of self-excitation in generators can aggravate the situation. When the generator is supplying capacitive current (due to the line capacitances), there is an upper limit on the current above which, the generator has overvoltage which cannot be limited by the Automatic Voltage Regulator (AVR). In such cases, it is necessary to ensure that there is sufficient generation capacity at the power station even while supplying low loads.

The protection against overvoltages is provided by surge arrestors (also called as lightning arrestors). The modern arrestors do not have spark gaps and use highly non-linear metal oxide resistors, (comprising primarily of zinc oxide). A typical disc of a surge arrestor, which conducts less than a milliampere of current at normal operating voltage, can carry currents of thousands of amperes at twice the voltage. This enables reduction of voltage margins and thus, insulation levels. However, gapless arrestors also have the duty of dissipating energy due to the high currents flowing in the disc during the voltage surge. The zinc oxide elements are connected in parallel to absorb the total energy in the surge.

6.5 Protective Relaying

The circuit breakers are equipped with trip coils which are energized by the operation of protective relays. In designing the relays, the following factors need to be considered

1. Reliability; 2. Selectivity; 3. Sensitivity.

A fault has to be cleared reliably to avoid failures or damages to the equipment. This implies that if the main relay fails to operate, there should be back up relay that should operate after some time lag. The relay should be sensitive to respond to the fault current even in the presence of fault resistance (which can limit the fault current). The selectivity implies that a relay should operate correctly when required. It should work to isolate the part of the system or equipment which it is supposed to protect against faults. Incidentally there could be hidden failures in the protection system, particularly since the faults occur rarely in a well designed system.

The relay operation is based on signals obtained from the current flowing through the breaker and the voltage at the location of the breaker. The current and voltage signals are derived from the instrument transformers - for current and voltage measurements. There are three or four types of relays in power systems.

1. Overcurrent relays using time delays.
2. Differential relays.
3. Directional comparison relays.
4. Distance relays.

For protection against faults in a radial distribution system, (see Fig. 6.1), over current relays are used.

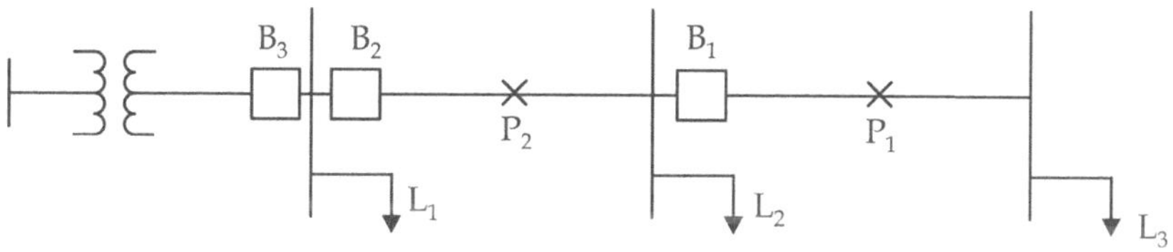

Fig. 6.1 A radial distribution system

If there is fault at P_1, breaker B_1 should operate (and not breaker B_2). If there is fault at P_2, breaker B_2 should operate and there is no need for the operation of B_1 as the tripping of B_2 will interrupt supply to loads L_2 and L_3. The co-ordination between the relays at B_1 and B_2 is achieved by having an inverse characteristic for the over current relay as shown in Fig. 6.2.

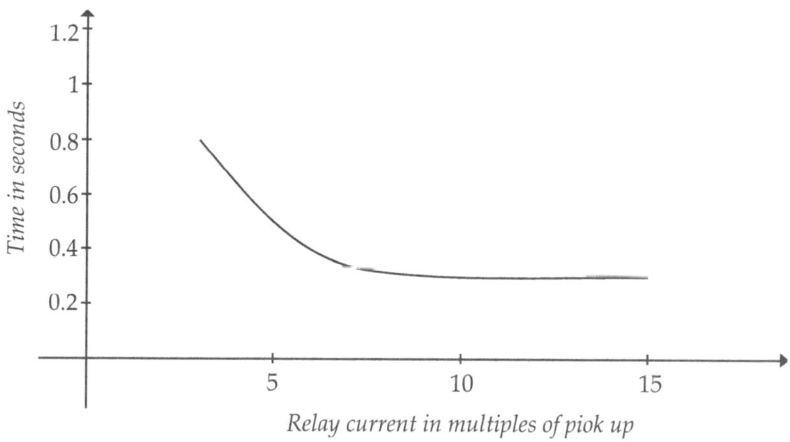

Fig. 6.2 Overcurrent relay characteristics

The time delay for the pick up of the relay is reduced as the fault current increases. The fault current at P_1 is less than the fault current at P_2. The breaker B_3 protects against faults in the zone between B_3 and B_2. It should be noted that breakers B_2 and B_3 also act as backup protection against fault at P_1. Similarly, breaker B_3 provides backup protection against fault at P_2. The coordination time interval is the time interval between the primary and remote backup protection devices. It is the difference between the time taken for backup relaying operation and the time when the circuit breaker clears the fault under primary relaying. The typical coordination time interval varies from 0.2 to 0.5 sec.

Differential relays are used for protection of generators, buses and transformers. Fig 6.3 illustrates the principle of operation of these relays.

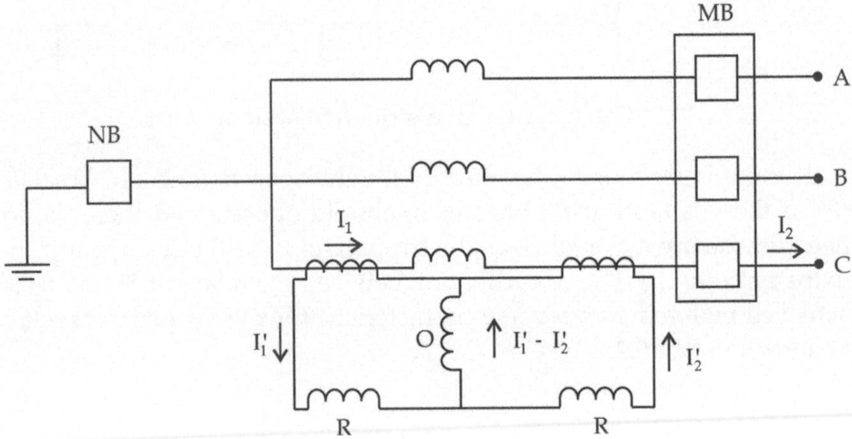

MB – Main Breaker; NB – Neutral Breaker

Fig. 6.3 A differential relay

Here, the protection of one phase winding of a generator is shown. If there is an internal fault, the current in the operating coil (O) is nonzero as $I_1' \neq I_2'$. The force exerted by the operating coil exceeds that exerted by the restraining coils (R). This will open the main breaker (MB) with contacts in all three phases. The neutral breaker (NB) and the field breaker (not shown) are also operated for faults in any one or more phases of the generator. For transformer protection, the current transformer ratios are suitably chosen to account for the turns- ratio of the transformer.

The distance relays utilize the measurements of the voltage and current phasors to compute the impedance $\left(\frac{\bar{V}}{\bar{I}}\right)$. The relay characteristic is based on depicting the region of operation in the R - X plane (where R + jX is the impedance). For example, a typical region of *mho* relay which operates only when X > 0 is shown in Fig 6.4.

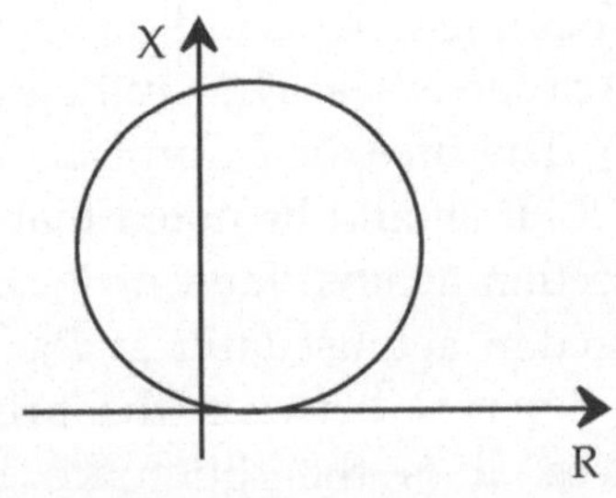

Fig. 6.4 Characteristics of a *mho* relay

The directional comparison relays are used to determine whether the fault is in the primary zone of protection or outside. For example, consider a line with two sections AB and BC as shown in Fig. 6.5.

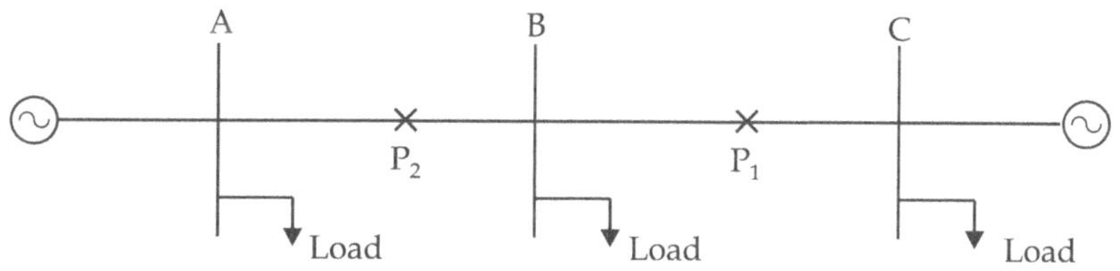

Fig. 6.5 A line with two sections

The fault currents flowing in the line AB from A and B for a fault at P_1 (in section BC) have same phases. On the other hand, the phase of current flowing from B (in section AB) reverses if the fault is at P_2 (in section AB). Generally in a transmission network, the fault current can flow from more than one source and it is necessary to trip two breakers, connected at the ends of the line section, to isolate a line section. Directional comparison relays can be used to block the operation of the breakers (connected at A and B) if the fault is at P_1. Thus, only the line section in which the fault occurs, will be isolated.

Fault Current Limiters (FCL)

As the power demand increases and the generation has to match the requirement by additional supply or system interconnection, the fault current also increases. It may not be feasible to replace existing breakers by new breakers with higher rating, particularly in urban areas (where the space constraints also play a role). In such cases, it becomes essential to limit the fault currents. There are two approaches to the design of FCL, one using high temperature superconductors (HTS), the other using solid state current limiter (SSCL). In the former design, a HTS conductor used in series provides zero resistance (acting as a superconductor) for the normal currents and inserts resistance when large fault currents flow (quenching the superconductor and making it normal, i.e lose its superconducting property). In another type of FCL design, a GTO (gate-turn-off thyristor) is used to break the circuit, however, the fault current continues to flow in a parallel HRC (High Rupturing Capacity) fuse which can limit the fault current within a millisecond(by inserting a resistance). Such FCLs are used for systems with rated voltages from 0.75 to 40 kV, with rated current up to 5 kA and breaking currents from 140 kA to 210 kA. The applications include bus ties, feeders, interfacing of renewable generation to the grid and cable protection.

6.6 System Collapse and Blackouts

So far, we discussed protection against the faults that can occur in transmission networks and other equipment. For generators, apart from protection against short circuits, there is need for protection against loss of field, over-excitation, loss of coolant, end winding heating etc. However, a major issue is the loss of synchronism (also called "out of step operation") caused by faults in the system that are not cleared fast enough (there is a critical clearing time for each fault depending on its nature and severity). For example, three phase faults (when all three phases are shorted) are rare, however cause maximum disturbance. When the clearing time (of the fault) exceeds the critical clearing time, the system can fall apart due to operation of individual protection systems. The generators and lines can trip based on their individual protection characteristics. Note that the primary objective of the individual protection scheme is to prevent damage to the equipment. If one generator trips and the remaining generation is inadequate to meet the total load, the frequency starts falling and it is likely that other generators also trip, one by one as they are overloaded and their individual protection system operate to trip them. Similarly, when a major transmission line trips and it cannot be reconnected quickly, it is possible that other lines start tripping due to overload. The cascading outage of lines can result in uncontrolled islands. It is possible that in a particular island, generation exceeds the load, resulting in frequency rise, whereas in another island, load is more than the available generation resulting in frequency decay. In either case, there will be extensive blackout and it may take hours or even days to restore power to all the customers.

When there is an imbalance between the load and generation, two possibilities can arise. Assuming that load exceeds generation, there can be voltage collapse or frequency collapse. The first event can occur when the system does not have enough reactive power reserves and the load characteristics demand constant power resulting in reactive power imbalance. Low frequency and/or voltage actuated load shedding can help preserve system integrity and avoid blackouts. If there is loss of synchronism and tripping of lines due to out of step relays, controlled system separation (instead of uncontrolled) can be initiated to prevent blackouts. This is termed as adaptive out-of -step protection as the lines that need to be tripped (to implement controlled system separation) can vary with the type of fault. The basic idea is that in the islands created, generation and load match, and the lines interconnecting the islands can be reclosed when the disturbance is over and steady state can be established. Even if load shedding is required, the quantum will be minimal and the system can be restored to the normal state in a short time.

Loss of synchronism can happen when (a) there is a large disturbance (which is rare) or (b) with increase in the loading of lines, spontaneous small oscillations in the power flow in lines are initiated due to instability of the equilibrium. When these oscillations grow sufficiently in magnitude, the generators will lose synchronism and the system can collapse. The second cause for instability is more troublesome as the system operators are not aware of the impending problem. Since adding additional lines to relieve stress on the existing network is not feasible, a practical solution would be to install FACTS controllers in the critical lines to avoid emergency situations. This will be discussed in the next chapter.

6.7 An Interesting Analogy

A well known Economist, Professor Jeffrey D.Sachs from Columbia University, USA wrote in Scientific American Magazine (December, 2008) in an article entitled 'Blackouts and Cascading Failures of Global Markets', "The global economic crisis is akin to a power blackout. In both cases, a disturbance in one part of complex "tightly coupled" system results in a cascade of failures through an entire network. In the case of a power blackout, a single downed power line or transient overload causes power to be shunted to another part of the grid, which in turn leads to new overloads, more shunting and ultimately to a cascade of failures that pushes a region into darkness. Similarly, in the current economic crisis, a U.S. banking emergency caused by worsening U.S. market forces has sent shock waves through the world's financial system, causing a global banking crisis that now threatens to lead to a global economic downturn."

Before we discuss the solutions for both crises, let us look at the costs of two recent crises in U.S- the power blackout in 2003 (14th August) and the 2008 economic crisis and their effects.

The power blackout in 2003 affected about 50 million people in Ontario, Canada and large portions of Midwest and Northeastern U.S. About 61,800 MW of load was lost and restoration took up to 4 days in some parts of U.S. In some parts of Ontario, there were rolling blackouts lasting one week. Total economic loss is estimated to vary from 6 to 12 billion dollars. In contrast, the economic collapse of 2008 led to

1. Shrinkage of GDP.
2. Increase in unemployment: According to ILO, 20 million jobs were lost by end of 2009.
3. Financial Markets: By early, 2009, 45 % of global wealth has been destroyed in less than a year and a half. In general, all types of economic activity are affected.
4. Political instability.
5. It is still not clear when the recession will be over and employment increase. In Southern Europe, even by end of 2013, the unemployment among young people is excessive. In India, the manufacturing activity has suffered. The currency has depreciated and inflation has affected all economic activity in addition to household budgets.

The interconnection of utilities by tie lines was motivated by economic benefits of reduced generation reserves, sharing of generation mixes, exploiting load diversity and economy of scale by introducing large generating units. However, the 1965 November blackout demonstrated that the tie lines can spread the crisis by interconnection. An effective measure for dealing with the emergency is to decouple the system by controlled system separation until the system can be restored to the original state of normalcy. It would be interesting to investigate whether similar action is feasible in economic systems. It may be possible to compare business cycles with power oscillations in transmission lines which can disturb system stability. However, it is feasible to damp power oscillations by effective control actions using Power System Stabilizers (in generating stations) and Power/Reactive Power oscillation dampers (POD) in FACTS and HVDC controllers. One can pose the question: can economic oscillations be damped?

In recent times, based on the GPS technology, efforts are in progress to design a self-healing grid using WAMS (Wide Area Measurement Systems). Although technology is available, suitable analytics are yet to be applied for the creation of self-healing grids. The concept of SPEF (Structure Preserving Energy Functions) can help in devising effective and optimal control strategies (Padiyar 2013).

CHAPTER 7

Technological Solutions and Smart Grid

"Technology is dominated by two types of people, those who understand what they do not manage, and those who manage what they do not understand"

Putt's Law

7.1 Introduction

It was mentioned in chapter 1 that the power demand at a load centre has to be met 'just in time' (at speeds approaching that of the light) from the generating stations that may be located hundreds of kilometers away. The power is to be delivered using high voltage transmission network and low voltage distribution feeders. The system can collapse if

1. the generation is inadequate to meet the load demand and the system frequency decays.
2. in addition to the power demand, there are reactive power constraints, the voltages at the load buses (nodes) can collapse leading to voltage instability and power blackouts.
3. the generators connected in the system lose synchronism due to a sudden large disturbance such as tripping of a line or lines following a major disturbance such as faults. The loss of synchronism (we also say that a generator (or group of generators) go out of step with respect to the remaining generators) requires the disconnection of the generator(s) that goes out of step. The problem is termed as transient (angle) instability.

4. when the transmission network is weak, at peak loads, the relative generator rotor angles tend to be high. In such cases, even a small disturbance (involving an increment in the load demand) can lead to oscillations in the generator rotors. When the oscillations increase in amplitude due to instability of the equilibrium, the generators can go out of step. This was termed as dynamic or small signal angle instability which is difficult to detect.

The problem of small, low frequency (0.2 to 2.0 Hz) oscillations was first observed in the mid sixties when high response static excitation systems with high gain automatic voltage regulators (AVR) were introduced. Such excitation systems with high gain AVRs were considered essential to overcome the problem of transient instability with the introduction of modern turbo generators. The problem of undamped low frequency oscillation is tackled by suitable design of power system stabilizers (PSS) that are auxiliary controllers whose output results in modulation of the generator voltages to damp the rotor oscillations. However, the design of PSS in a large interconnected system which can have several dominant modes of oscillation is still a challenging task.

The problem of voltage collapse due to instability was first experienced in the late seventies in France and subsequently also observed in Japan, U.S.A and other countries. A major factor in the voltage collapse is the failure of voltage regulation in generators and the action of on-load tap changers (OLTC) at the load bus which tend to regulate the bus voltage and try to restore the load.

The load restoration also occurs inherently in induction motors following a dip in the voltage. It is to be noted that the collapse of voltage at the load buses results in black out although the generators remain in synchronism and the frequency rises.

7.2 Protection against System Collapse

The frequency variation due to imbalance in the load and generation is limited by the inherent nature of the motor loads that draw power as a function of the bus frequency. With reduction in frequency, the power demand also decreases. Similarly, the voltage dependant characteristic of the heating and lighting loads can aid in voltage regulation (the power demand decreases as the voltage falls). This property is used to reduce voltage (called as brown-outs) in times of shortage of spinning reserve to avoid power interruptions. The turbine governor characteristic is also designed to limit the frequency variation and proper sharing of the power generation among generators connected in a power station.

However, when the generation is inadequate to supply the load, frequency actuated relays are used for load shedding to prevent an emergency. It may also be necessary to provide relays that will initiate load shedding at low voltages to prevent voltage collapse. The load shedding can take into account the priorities of various loads to avoid interrupting power to hospitals, railways etc.

Application of High Power Electronic Controllers

The high voltage DC (HVDC) transmission technology underwent a major change when thyristor (a power semiconductor device) based power converters were applied in early seventies instead of mercury arc valves. Initially, the current ratings of thyristors were inadequate and they had to be connected in parallel (in addition to series connection to provide the required voltage level) to provide the required current rating. However, the current ratings have now increased to supply over 3000 A of DC current (continuous rating). The voltage rating has also increased, and the number of series connected devices in a converter arm has decreased for a specified voltage.

The reliability of the microprocessor based converter control has made it possible to eliminate the need for HVDC breakers for clearing faults in a HVDC line (it is to be noted that HVDC breakers are costly and take longer time to clear a fault as there is no inherent current zero occurring in a DC circuit. The breaker has to develop a counter voltage of sufficient magnitude to drive the DC current to zero. The time required to clear the fault depends on the inductance in the circuit).

The rapid control action of the converter control makes it possible to control the power flow in a fast manner that can improve the stability of the associated AC system. The converter control can be used to regulate the power flow and also modulate (in response to the deviations in frequency) as required in order to damp low frequency oscillations. The converter control can be used also to regulate frequency in the AC system. The successful implementation of the thyristor based converters has spurred the application of thyristor based controllers in AC systems. In late seventies, the thyristor controlled reactors (TCR) and thyristor switched capacitors (TSC) were used to develop Static Var Compensators (SVC) that can provide variable and fast acting reactive power compensation to regulate bus voltage at specific locations. The application of SVC helps to improve AC system stability by improving the voltage profile in the network and also damp power oscillations by modulation of the injected reactive power. The first major application of SVC was in the 735 kV lines of Hydro Quebec in Canada.

7.3 Flexible AC Transmission System (FACTS)

In 1988, Dr. Narain Hingorani of EPRI (Electrical Power Research Institute) in U.S.A proposed the concept of FACTS to improve the system operation. This involves the application of high power electronic controllers in AC networks to regulate voltages and power flows in lines both under steady state and dynamic conditions. The benefits of FACTS are:

1. They contribute to optimal system operation by reducing power losses and improving voltage profile.
2. The power flow in critical lines can be enhanced as the operating margins can be reduced due to fast controllability. In general, the power carrying capacity of transmission lines can be enhanced up to thermal limit (imposed by the current carrying capacity of the line conductors). This improves ATC (Available Transfer Capacity) of lines, thereby reducing congestion in AC networks.
3. The problems of voltage and angle instability can be overcome both under small and large disturbances.

The SVC can be considered as the first FACTS controller, connected in shunt to improve voltage profile. The second major FACTS controller is Thyristor Controlled Series Capacitor (TCSC) using Thyristor devices. Here, a series connected capacitor is connected in parallel with a TCR to provide variable series compensation. Prototype TCSCs were introduced in early nineties. Now the technology has matured and there are several TCSC installations in the world including China and India.

The high power electronic controllers applied in transmission lines have evolved from the technology developed for adjustable speed devices where Current Source Converters (CSC) using thyristor devices have been replaced by Voltage Source Converters (VSC) using IGBT (Insulated Gate Bipolar Transistor) devices. Thus, it was natural to improve SVC (which is basically a variable impedance type converter) by a novel compensation using VSC. Initially, Gate Turn Off (GTO) devices were used as high power IGBT devices were not available. Unfortunately, GTO devices are not suitable for series connection and have low switching frequency. The reactive power compensator using VSC is named as STATCOM (STATic Synchronous COMpensator). It was originally named as STATCON (Static Condenser) because its characteristics are similar to a rotating synchronous condenser (which is an unloaded synchronous motor designed for reactive power compensation). A STATCOM is superior in performance compared to a SVC. It requires a smaller footprint and hence suitable for installation in urban areas.

A series connected VSC based FACTS controller is called as Static Synchronous Series Compensator (SSSC) and is an improvement on TCSC. Dr. Gyugyi and his co-workers at Westinghouse Company in U.S.A. have also developed two new FACTS controllers based on two or more VSC. (a) Unified Power Flow Controller (UPFC) which can control 3 parameters - power and reactive power flow in a line in addition to the bus voltage where a shunt and a series connected VSC are located. (b) Interline Power Flow Controller (IPFC) where two or more VSC connected in series in different lines that are typically connected to a bus where power is injected. The IPFC can regulate power and reactive power in all lines (where VSC is connected) except one where only reactive power flow is regulated.

It is possible to generalize the concept of multi converter FACTS controllers based on VSC. We can have the converters connected in shunt or series with their DC capacitors connected in parallel. Thus, we can have generalized UPFC that can regulate bus voltages in more than one location; power and reactive power in more than one line. Furthermore, flexibility is introduced when two or more converters can be connected in different configurations depending on requirement. A Convertible Static Compensator (CSC) with two converters has been installed in New York state in U.S. with two transformers (one connected in shunt and other in series) to operate with up to 10 different configurations.

7.4 Principle of Switching Control

In power electronic converters, the control of voltage and power is achieved by switching. As a simple example, see Fig. 7.1(a) where a DC voltage source is connected to a load (assumed to be a resistor) through a switch S. When the switch is closed, current and power flows from the source to the load. The current and power are interrupted when the switch is open. Assuming that the switch turns on and off in a cycle of fixed period (T), by varying the duration (T_{ON}) of the switch when it is on, the power flow to a load can be varied. The variation of voltage across the load is shown in Fig. 7.1 (b). The average voltage V is ET_{ON}/T. The ripple content in the load voltage can be removed by a filter. Unlike a mechanical switch which has limitations in speed and the frequency of operations due to the movement of contacts and arcing problems, a power semiconductor switch such as IGBT can handle frequency of operation up to 2000 Hz. Also, the switching losses are low (not zero as in an ideal switch). Note that in an ideal switch, the voltage (across the switch) is zero when closed and current (through it) is zero when open and hence there is no power consumption. In a thyristor, there is an on-state voltage drop which also depends on the current flow.

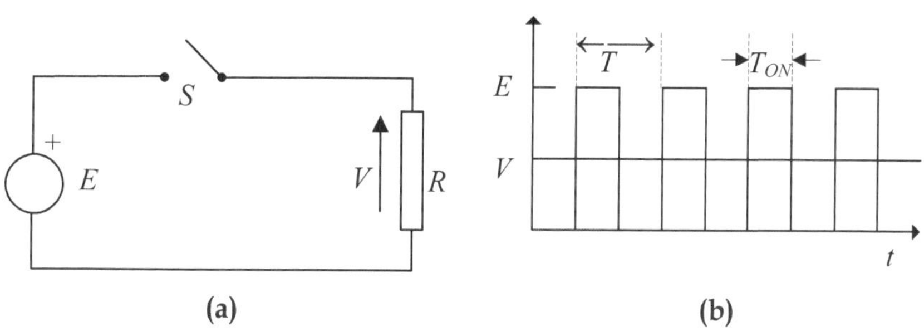

Fig. 7.1 Illustration of switching control: **(a)** A DC voltage source supplying resistive load in series with a controllable switch **(b)** Voltage variation across the load

In the earlier days when power semiconductor switches were not available, rheostatic control was used (a variable resistance connected in series with the voltage source) and this consumed energy. The introduction of power semiconductor switches has resulted in significant energy savings which translates into reducing the carbon foot print which affects climate change and consequent global warming.

It is to be noted that it is possible to generate AC wave form by switching alternately (at twice the fundamental frequency) positive and negative DC voltages. The generation of pulses in each half of the fundamental frequency sine wave is called as pulse width modulation (PWM) for controlling the magnitude and phase of the fundamental sine wave. The pulses introduce high frequency harmonics which are filtered out.

7.5 Wide Area Measurement System (WAMS)

Whenever there are major disturbances that can threaten the system security, it is necessary to detect by simultaneous measurement over a wide area (covering the entire network spread over a large area), such that effective remedial actions (emergency control) can be taken to prevent (or at least limit the spread) blackouts. This would be the first step in the planning of a self- healing grid. This approach is based on the utilization of technological advances in communications, computers and control. A major development is the measurement of voltage and current phasors (the magnitudes and phase angles at different locations). Since phase angles are all referred to a common reference, precise time synchronization is required which is achieved with a global positioning satellite (GPS) system. GPS is a U.S. government sponsored program that provides worldwide position and time broadcasts.

A major prerequisite for optimal system operation is the availability of accurate time stamped measurements of various quantities that are of importance. For example, the transmission line rating in real time depends on the ambient temperature. In winter times the line can carry more power as the thermal limit is based on the maximum temperature of the aluminum conductor. If the ambient temperature can be accurately measured or estimated, then the line can be fully utilized at any given time (available transfer capacity can be maximized). This is known as dynamic rating. The ambient temperature can also be estimated based on the conductor sag. The time of the day metering is considered necessary when electrical energy is priced depending on the demand. In such cases, a customer should have price signals so that he or she can minimize the energy bill either by conservation or use of alternative sources (assuming there is a choice).

7.6 GRID 2030-A National Vision

United States Department of Energy (DoE) Office of Electric Transmission and Distribution published a document "GRID 2030: A national vision for Electricity's Second 100 years" in July 2003 with the objective of transforming the Grid to revolutionize electric power in North America. This was the outcome of a meeting held on April 2-3, 2003 where 65 senior executives representing the electrical utility industry, equipment manufacturers, information technology providers, Federal and state government agencies, interest groups, universities and National Laboratories met to discuss the future of North America's electric system.

The vision of "Grid 2030" is that "It energizes a competitive North American market place for electricity. It connects everyone to abundant, affordable, clean, efficient, and reliable electric power anytime, anywhere. It provides the best and most secure electric services available in the world."

They found that although the America's electric system is "the supreme engineering achievement of the 20th century", it is aging, inefficient and congested and incapable of meeting the future energy needs of the Information Economy without operation changes and substantial capital investment over next several decades".

The recommendations of this meeting were:

1. The "technology readiness" of critical electric systems needs to be accelerated, particularly for high-temperature superconducting cables and transformers, lower-cost electricity storage devices, standardized architectures and techniques for distributed intelligence and "smart" power systems, and cleaner power generation systems, including nuclear, clean coal, renewable, and distributed energy devices such as combined heat and power.
2. A breakthrough is needed to eliminate the "political log jam" and reduce the risks and uncertainties caused by today's regulatory framework. This includes clarifying intergovernmental jurisdiction, establishing "rules of the road" for workable competitive markets wherever they can be established, ensuring mechanisms for universal service and public purpose programs, and supporting a stable business climate that encourages long term investment.

3. Industry will be investing billions of dollars over the next several decades to replace electric power equipment. The economic life of this new equipment will last 40 years or more, so this turnover of the nation's capital stock of electric power assets needs to include the latest technologies to ensure clean, efficient, reliable, secure, and affordable electricity for generations to come. An expanded research, development, and deployment effort is paramount.
4. A logical next step is the collaborative development of a National Electric Delivery Technologies Roadmap. This technology Roadmap will be principally used for guiding public and private research, development, and demonstration programs.

The GRID 2030 document also gives details of how to achieve the vision and the grand challenges facing the task.

Discussion

The vision document "GRID 2030" prepared by the Department of Energy, U.S. is very ambitious and appears more like a wish list rather than a realistic plan for the development of the Smart Grid. It is not clear how much progress has been made towards achieving the vision. For example, there is not much action on the ground for the development of High Temperatures Superconductors (HTS) for application in generators, transformers and cables. China has installed a 33 kV HTS cable on an experimental basis. The real issue is the funding of research projects to achieve the vision. Incidentally, the Power Electronics Research and Development Program Plan (2011) by DOE, U.S.states "Power Electronics (PE) will play a critical role in transforming the current electric grid into the next-generation grid. Existing silicon-based PE devices enable electric grid functionalities such as fault-current limiting and converter devices. However, silicon (Si) based semiconductor technology cannot handle the power levels and switching frequencies required by next generation utility applications-hundreds of kilovolts(kV) blocking voltages at tens of kilohertz (kHz). PE devices based on wide band gap (WBG) semiconductor materials, such as silicon carbide (SiC), gallium nitride (GaN), and diamond could increase the reliability and efficiency of the next generation electric grid. These materials are capable of higher switching frequencies (kHz) and blocking voltages (upward of tens to hundreds of kV), while providing for lower switching losses, better thermal conductivities and the ability to withstand higher operating temperatures. A number of barriers and challenges exist in utilizing WBG PE devices to their full potential including identifying new device topologies for high-power grid applications, developing the ability to consistently deliver robust devices, and creating a cost-effective, high-volume manufacturing process. Material-specific barriers include:

GaN: producing bulk GaN requires reaching extremely high decomposition pressures and a high melting temperature.

SiC: defect density control and material availability remain issues, despite decades of research.

The goal is to develop by end of 2016, a 5 kV at 15 A GaN-Si (Gallium Nitride-on-Silicon) PE devices."

It is interesting to note that with the existing PE devices(such as IGBT) and using the topology of Modular Multilevel Converter (MMC), several HVDC links based on VSC technology are being introduced in Europe with capacities in the range of 1000 MW. The MMC topology does not require high switching speed and thus the converter losses are substantially reduced. Engineers have always made the best use of existing and developing technologies to implement cost effective solutions without compromising on reliability and performance.

7.7 What is a Smart Grid?

According to International Electrotechnical Commission (IEC), a "Smart Grid" is used today as a marketing term, rather than a technical definition. For this reason, there is no well defined and commonly accepted scope of what "Smart" is and what is not.

The general understanding is that the Smart Grid is the concept of modernizing the electric grid. According to U.S. Department of Energy, "Smart Grid generally refers to a class of technology people are using to bring utility electric delivery systems into the 21st century, using computer based remote control and automation. These systems are made possible by two-way communication technology and computer processing that has been used for decades in other industries." According to IEC, "The smart grid comprises of everything related to the electrical system in between any point of generation and any point of consumption. Through the addition of Smart Grid technologies, the grid becomes more flexible, interactive and is able to provide real time feedback. It is an electricity network that can intelligently integrate the actions of all users connected to it - generators, consumers and those that do both – in order to efficiently deliver sustainable, economic and secure electricity supplies. Smart technologies improve the observability and/or the controllability of the power system. Therby, Smart Grid technologies help to convert the power grid from a static infrastructure to be operated as designed, to a flexible, "living" infrastructure operated proactively."

Examples:

- Smart metering could significantly improve knowledge of what is happening in the distribution grid, which nowadays is operated rather blindly. For the transmission grid an improvement of the observability of system-wide dynamic phenomena is achieved by Wide Area Monitoring and System Integrity Protecting Schemes.
- HVDC and FACTS improve the controllability of the transmission grid. Both are actuators, e.g. to control the power flow. The controllability of the distribution grid is improved by load control and automated distribution switches.
- Common to most of the Smart Grid technologies is an increased use of communication and IT technologies, including an increased interaction and integration of formerly separated systems.

European Technology Platform for Smart Grid (2010) defines smart grid as follows:

"A Smart Grid is an electricity network that can intelligently integrate the actions of all users connected to it - generators, consumers and those that do both - order to efficiently deliver sustainable, economic and secure electricity supplies. A Smart Grid employs innovative products and services together with intelligent . monitoring control, communication and self-healing technologies to:

1. allow consumers to play a part in optimizing the operation of the system, and
2. provide consumers with greater information and choice of supply.

Smart Grid deployment must include not only technology, market and commercial considerations, environmental impact, regulatory framework, standardization usage, ICT (Information & Communication Technology) and migration strategy, but also societal requirements and governmental edicts."

Discussion

The smart grid (sometimes also called as intelligent grid) concept appears to have evolved after the failure of the deregulation/ restructuring in U.S. in reducing costs of electricity and at the same time impairing reliability and security. The U.S. government declared a 3 billion dollar grant for research and development in smart grid technology in 2009.

In the recent book titled "Smart Grid: Communication -Enabled Intelligence for the Electric Power Grid (2014) by Stephen F. Bush, it is mentioned "The power grid is very broad system comprised of many different components and technologies. Researchers focusing on a narrow aspect, such as developing smart meters or developing new types of demand-response(DR) mechanisms, sometimes inadvertently equate their areas with the sum total of the smart grid"..... This is similar to the story of an elephant and six blind men. "Each blind man equates the elephant with the part he can feel. The areas shown are:

1. Advanced metering infrastructure (AMI)- systems that measure, collect, and analyze energy usage;
2. Distribution automation (DA)-the extension of intelligent control over electrical power grid functions to distribution level and beyond;
3. Distributed generation (DG)-generating electricity from many smaller energy sources and microgrids;
4. Substation automation (SA)-automating electric power distribution and transmission substation;
5. Flexible alternating current transmission system (FACTS)-a power electronics-based system to enhance controllability and increase power transfer capability of the network; and
6. DR- systems that manage customer consumption of electricity in response to supply conditions."

7.8 Microgrids

According to Microgrid Exchange Group (2010), "A microgrid is a group of interconnected loads and distributed energy resources (DER) within clearly defined electrical boundaries that act as a single controllable entity with respect to the grid. A microgrid can connect and disconnect from the grid to enable it to operate in both grid-connected or island-mode.

For microgrid to work properly, an upstream switch must open (typically during an unacceptable power quality condition) and DER must be able to carry the load on the islanded section. This includes maintaining suitable voltage and frequency levels for all islanded loads. Depending on switch technology, momentary interruptions may occur during transfer from grid-connected to island mode. In this case, the DER should be able to restart and pick up island loads (and also handle inrush during "starting" of the island).The DER must be able to supply the real and reactive power requirements during islanded operation and to sense if a fault current occurred downstream of the switch location. When power is restored on the utility side, the switch must not close unless the utility and "island" are synchronized.

The objectives of the microgrids are to provide (a) higher reliability of service and (b) better power quality to the end customers. Microgrids can also provide additional benefits to the local utility by providing dispatchable power for use during peak power conditions and alleviating or postponing distribution system upgrades.

Microgrids consist of several basic technologies, such as Distributed Generation (DG), Distributed Storage (DS), interconnection switches and control systems. DG units are small sources of energy(typically, photovoltaic(PV), wind, fuel cells, microturbines and reciprocating internal combustion engines with generators located at or near the point of use. These systems may be powered by either fossil or renewable fuels. Distributed storage (DS) provides a bridge in meeting the power and energy requirements of the microgrid. Storage capacity can be categorized in terms of the energy density requirements (for medium and long term needs) or power density requirements (for short and very short term needs). DS enhances the overall performance of the microgrid in 3 ways: 1. it stabilizes and permits DG units to run at a constant and stable output, despite load fluctuations, 2. it provides ride-through capability when there are dynamic variations of primary energy (such as solar, wind and hydropower sources) and 3. it permits DG to seamlessly operate as a dispatchable unit.

There are several forms of energy storage such as batteries, super-capacitors and flywheels. These will be discussed in the next chapter.

Interconnection switches are designed to meet grid interconnection standards(IEEE 1547 and UL 1741 for North America) to minimize custom engineering .The controls are also designed to be technology neutral and can be used with a circuit breaker as well as static switches using thyristors or IGBTs. The control of frequency during islanded mode is more problematic than that during grid connected mode of operation. The control of voltage is related to the generation of reactive power locally and tends to be independent of the mode of operation of the microgrid.

Around the world, there are several active experiments in the microgrid area covering various technologies. For example, in U.S., Consortium for Electric Reliability Solutions (CERTS) has a test bed in Columbus, Ohio. Tests are also being conducted in Canada, Japan and Europe.

In summary, distribution grids are being transformed from passive to active networks in the sense that decision making and control is distributed and power flows are bidirectional from a load node (also acting as a generation node at times). It is to be noted that Smart Grid concepts are being primarily applied to microgrids operating at low or medium voltages.

7.9 Role of ICT in the Development of Smart Grids

The decision makers expect that those connected with information and communication technologies (ICT) have the major role to play rather than electrical engineers who have been planning, designing and operating power systems to meet the load demand at affordable costs at the desired reliability and quality (of power). However, there are two issues that must be considered before entrusting the complete responsibility to ICT practitioners who have little knowledge about electricity supply systems. The first one is about the increasing electric demand in the ICT eco system (consisting of Data centres, Broadband wired and wireless communication networks, end-use devices from PCs to tablets and smart phones to digital TV, the manufacturing facilities producing ICT hardware). The world's ICT ecosystem uses about 1500 TWh of electricity annually, equal to all the electrical generation of Japan and Germany combined and it now approaches 10 % of world electricity generation(according to a recent report titled "The Cloud Begins with Coal-Big Data, Big Networks, Big Infrastructure and Big Power" , 2013).

Although charging up a single tablet or smart phone requires negligible amount of electricity, using either to watch an hour of video weekly, consumes annually more electricity in the remote networks than two refrigerators use in a year. And as the world continues to electrify, migrating to one refrigerator per household, it also evolves towards several smart phones and equivalents per person. Also, the inherent nature of the mobile Internet, a key feature of the emergent Cloud architecture, requires more energy than do wired networks. The trends are promoting faster growth in ICT energy use. For example, hourly internet traffic will soon exceed the annual traffic of the year 2000! The demand for data and bandwidth and the associated infrastructure are growing rapidly not just to enable new consumer products and video, but also to drive revolutions in everything from healthcare to cars and from factories to farms.

Electricity Demand in Data Centres

There are tens of thousands of data centres around the world. A typical data centre costs roughly $ 7 million per megawatt to build, and another $ 9 million per megawatt for the cost of electricity over the facility's 10-year operating life, assuming low cost power. Thus, for example, a single 50 MW enterprise data centre sited in Iowa (70 % coal, 25 % wind) instead of higher cost California (no coal), saves $ 350 million in electricity charges over the life of that single data centre. But 80 % of the global ICT electricity use is highly dispersed. Thus the cost and availability of electricity for the Cloud is dominated by the same realities as for the society at large-obtaining electricity at the highest availability and the lowest cost. The global ICT electricity use is growing and the implications of this trend are commented by the National Academy of Sciences as "As the use of the Internet continues to grow and massive computing facilities are demanding that performance keep doubling, devoting corresponding increases in the nation's electrical energy capacity to computing may become too expensive."

Cyber Security

The cyber security is the second major issue. In a recent article in IEEE Spectrum (November 2013), Karl Rauscher states "In the 21st century, just about every thing is vulnerable to cyber attack. A hit on the bank or a stock exchange would cause uproar in the financial sector; a strike on electrical grid could shut down a city. And the consequences of an attack could be far more dire than mere inconvenience. If hackers disrupted operations at a nuclear power facility, they could trigger a meltdown. An attack on a hospital could leave doctors scrambling in the dark, machines failing, and patients dying in the beds. Such scenarios are becoming ever more plausible." The smart grid envisages power managers at homes that will turn on or off power appliances (such as heaters) automatically in response to pricing signals sent by the system operator. It is not obvious what preventive measures are to be used to provide security against attacks by viruses, malware etc.

Conclusions

The smart grid, as discussed extensively in publications, conferences and research projects, specifically pertain to distribution systems operating at low or medium voltages. The concept of technically savvy microgrids appears to be the platform for the application of smart grid technologies. However, it is not obvious who will fund these and what are the costs involved.

Furthermore the application of smart grid technology would depend on economic viability and customer needs which vary from place to place. In India, for example, where customers witness frequent power interruptions, even in metros, the communication from the DISCOMs prior to power cuts would be a welcome proposition. Also, the customers who practice energy conservation need to be encouraged, if it is feasible to segregate the customers based on their consumption patterns.

CHAPTER 8

Renewable Power and Energy Storage

"We tend to overestimate the effect of a technology in the short run and underestimate the effect in the long run"

Amara's Law

8.1 Introduction

Traditionally, electric power generation is based on the utilization of chemical energy in fossil fuels. These are primarily coal, oil and natural gas. Coal has been used since early part of 19th century, whereas oil began to be used since beginning of 20th century. The gas is being used since 1960. Coal is the end product of a natural process of decomposition of vegetable matter buried in swamps and out of contact with oxygen, for thousands of years. The coal includes a wide variety of solid fuels such as peat, lignite, bituminous and anthracite coal. The availability of fossil fuels is limited. In addition, they cause pollution and give rise to carbon emission that can result in climate change. The hydro-electric and nuclear power plants do not have this problem but tend to be costly, have long gestation period and different set of problems. For example, large hydro electric projects involve construction of big dams and nuclear power has safety issues. The latest scare was caused by the leakage of radioactive water from Fukushima nuclear plant in Japan. This has led to the development of safer energy sources such as wind and solar which are renewable. There are other forms of renewable energy such as tidal, wave and geothermal.

The wind energy is growing at the rate of 30% per year and the installed capacity was 282,482 MW at the end of 2012. The installations of photovoltaic cells (utilizing solar energy) exceed 100,000 MW.

The role of nuclear energy in the power development is described in Appendix C.

8.2 Solar Energy

The power of raw sunshine at midday on a cloudless day is 1000 W per square meter at equator. The total power from the sun intercepted by the earth is approximately 1.8×10^{11} MW which is thousands of times the present consumption. However, the solar energy is a dilute source with total radiation over the day does not exceed 7kWh/m^2. This implies need for large collecting areas and higher costs. The additional problem is large variation in the availability of the solar energy in a daily cycle (of day and night). The seasonal variations also occur due to earth's orbit around the sun. Variations also occur due to the local weather conditions. Hence, it becomes necessary to store the energy collected (when the sun is shining) to be used when the solar energy is not available. This feature adds to the cost of solar energy.

The solar energy can be utilized in the following three ways:-

1. Solar thermal: Using sunshine for direct heating of buildings or water. In desert locations, with concentrated solar energy, it is feasible to generate steam and use it for electric power generation.
2. Solar photo voltaic: For generating electricity.
3. Solar biomass: Using trees, bacteria, algae, corn, soya beans or oil seeds to make energy fuels.

Photo Voltaic (PV) Conversion

Direct conversion of solar energy into electrical energy occurs in solar cells using semiconductor materials such as silicon, gallium arsenide etc. Solar cells or PV panels are best installed on roof tops. The maximum efficiency of a solar cell is about 20%. Thus, 1 m^2 of solar panel can generate a maximum of 214 watts. This power is available only during sunlight hours.

The working of solar cell involves two steps:

1. Creation of pairs of positive and negative charges (called electron-hole pairs) in the cell by absorbed solar radiation.
2. Separation of positive and negative charges by a potential gradient within a cell.

The first step requires that the cell must be made of a material which can absorb the energy associated with the photons in the sunlight. The energy *E* (in electron-Volts (*eV*)) is related to the wave length of the light by

$$E = 1.24/\lambda$$

where λ is the wave length in micrometers (μm).

In a semiconductor, the electrons occupy either the valence band or the conduction band. The valence band has electrons at lower energy level and is fully occupied whereas the conduction band has electrons at higher energy level and is not fully occupied. The difference in the energy between the two bands is termed as band gap energy.

Photons of sunlight having energy levels greater than the band gap energy are absorbed in the cell material and excite some of the electrons to jump from valence band to the conduction band leaving behind holes in the valence band. Thus, electron-hole pairs are created.

The electrons in the conduction band and holes in the valence band are mobile and can flow through an external circuit if a potential gradient exists in the cell. In silicon material, this is obtained by making the cell as sandwich of two types of silicon-*p* and *n* types. The potential gradient occurs at the junction of the *p* and *n* type, which causes a direct electric current to flow in the circuit supplying the load.

Biomass

The plant matter created by the process of photosynthesis (which derives its energy from the solar radiation) is called biomass. In the process of photosynthesis, water and carbon dioxide are converted into organic material. The term 'biomass' includes all plant life-trees, agricultural plants, bush, grass, algae and their residues after processing.

The electrical power can be produced from energy plantations which produce fuel for the boilers in a conventional power plant. This technology is well established in USA and Europe, with plant outputs in the range of 5 to 50 MW. These power plants use wood, wood waste and agricultural waste. Assuming a photosynthesis efficiency of 1%, it is estimated that a 1000 MW plant may require an area of 1000 square kilometers for the energy plantation.

Biomass can also be used indirectly by converting it into a usable fuel in solid, liquid or gaseous form. There are two forms of conversion- 1. thermochemical and 2. biological. The first involves destructive distillation, pyrolysis and gasification. The biological conversion process includes processes such as fermentation.

8.3 Wind Power

The wind power has been used for centuries in sailing ships, for pumping water and grinding grain. The power contained in the wind is obtained from the kinetic energy of the flowing air mass per second. It is proportional to the cube of the wind velocity. The movement of the air molecules is not unidirectional and hence the velocity of the wind is average velocity in a specific direction. There is a theoretical maximum limit of efficiency of conversion which is 16/27 (called the Betz limit). It is to be noted that wind turbines must operate on the utilization of aerodynamic forces rather than thrust (which involves striking a blocking surface). The maximum power extracted from a thrust operated turbine is only 8/27 times the power contained in the wind.

The wind farm consists of a group of wind turbines in the same location. In a large wind farm (rated up to 1000 MW) several hundred, horizontal axis wind turbines, each with 3 blades attached to a nacelle on the top of a tall tubular tower are employed. The tower supports the complete machinery including the blades, gear box, the generator and the control equipment and hence requires high strength. Since the optimum speed of rotation of the generator is much higher than the optimum speed of the turbine blades, a gear box is inserted in the torque transmission system. There is an optimum gear ratio (20-30) that gives maximum efficiency for a designed speed. For a given wind speed, there is an optimum turbine speed at which the extracted wind power is maximum.

Wind turbines require control systems to maximize the power extracted in the specified range of wind speeds, to protect the generator and power electronic equipment from overloading during high wind speeds. At wind gusts, the machine has to be stalled. Typically, the range of wind speeds for normal operation is 9 to 16 m/s. There are three different types of controls.

1. *Pitch angle control*: This enables the turbine to have high efficiency by continuously aligning the blade in the direction of the wind. When the wind speed exceeds the rated speed, the blades are gradually turned about the longitudinal axis and out of the wind to increase the pitch angle. This reduces the output power of the turbine. When the wind speed exceeds the safe limit, the pitch angle is changed to reduce the power output to zero and the turbine shifts to the stall mode.
2. *Stall control*: In constant speed wind power systems using induction generators, the rotor speed varies in a narrow range (1 to 4%). As the wind speed increases, the lift force decreases for the rotor running at near constant speed. The rotor blade profile can be suitably shaped to create turbulence in the rotor blade side not facing the wind, thus resulting in reduced efficiency. This is called as passive stall control.

   In active stall control, the blades are rotated by a few degrees in the direction opposite to that in a pitch controlled machine. This helps to keep the output power at its rated value at high wind speed.
3. *Power electronic control*: By introducing a power electronic interface between the wind power generator and the grid, it is possible to regulate the power output by power electronic control.

There are three types of wind generators used in wind power systems.

1. Squirrel cage induction generators with shunt connected capacitor banks.
2. Wound rotor induction generators with variable rotor resistances or with power electronic control. In the latter case, it is called as doubly fed induction generator (DFIG).
3. Permanent magnet synchronous generator with frequency controller. It is to be noted that a DC link between the wind power generator and the grid acts as frequency converter.

Off Shore Wind Farms

At sea, winds are stronger and steadier than on land and hence, offshore wind farms produce higher energy than onshore wind farms.

According to David McKay (2009), the offshore wind power is 50 % more than the onshore value in U.K.

In the recent times, with emphasis on wind power, there are significant developments taking place, particularly in Europe, to exploit this power. The power is transmitted to onshore grid using VSC-HVDC transmission which has several advantages over LCC-HVDC based on the thyristor devices.

8.4 Hydro Electric Power

The hydro power based on flow of river water has been developed for a long time and provides renewable energy that forms 16% of the total electrical energy generation (globally). The hydro power can be classified into following categories.

1. Storage reservoirs with unidirectional flow from the reservoir to the power house located at the lower level.
2. Run of the river plants in which the water in the river is diverted by a retaining wall to flow through water turbines designed for low heads, but large quantities of water.
3. Pumped storage plants in which the water is pumped from the downstream side of the dam to the reservoir to raise its storage capacity during the periods of low loads (usually at nights). At peak load conditions, the hydro-turbines in the power plant drive the synchronous machines to generate power and supply to the grid. Sometimes, a separate pump is used to fill the reservoir. It is obvious, that these plants can be used for energy storage for leveling of loads or shaving the peaks.

8.5 Wave Energy

The wave energy is the result of interaction of the winds with the ocean surface. It is one of the indirect ways of utilizing the solar energy. The available wave power depends on the amplitude and frequency of the waves. It is estimated that about 10 kW of power is available in the ocean waves for every meter of the wave front and the total amount of the wave power is of the order of 2×10^6 *MW*. However, wave energy is difficult to collect due to wide fluctuations in the amplitude and frequency of the waves. A few systems have been built in UK and Japan along the shore at suitable locations.

There are several wave energy conversion systems. The most common is the oscillating water column system. Here, air is trapped between two columns of water in an inverted U-tube of large dimension which moves due to the pressure difference created by the rising and falling wave external to the tube. An air turbine is placed in the air passage which drives an electric generator.

8.6 Ocean Thermal Energy Conversion (OTEC)

This utilizes the temperature difference which exists naturally between the upper and lower layers of water in the ocean (due to solar energy). The difference is around 20 °C and occurs over a few hundred meters. The energy conversion system may be a closed Rankine cycle or open Claude cycle. The closed cycle system uses a low boiling point working fluid like ammonia, propane etc. Warm surface sea water is pumped through an evaporator where the working fluid in the form of a high pressure liquid absorbs heat and is converted into vapour. The vapour drives a turbine generator producing electricity. Then the vapour is condensed in a condenser using cold water brought up from the ocean depths through a cold water pipe. The condensed liquid is pressurized before it passes into the evaporator again, thereby closing the cycle. The open cycle system uses sea water as the working fluid. The warm, surface sea water is degasified and then evaporated in a flash evaporator under vacuum, thereby producing low pressure steam. The steam is expanded in a turbine which is coupled to a generator.

The efficiency of energy conversion in OTEC system is very low (<5%), however, the solar energy is free and the system may be economically viable.

8.7 Tidal Power

The gravitational and earth's rotational forces exerted on the sea water, give rise to high tides. In roughly 25 hours, there are two high tides and two low tides. Tidal power generation uses periods of high tides to fill a reservoir behind an embankment or barrage and during low tide periods, the stored water is used to run water turbines driving electric generators.

The highest amplitude for the high tide occurs during full moon and new moon days; these are known as spring tides. The height decreases as days progress towards the quarters and are lowest on days of first and third quarters when the sun is on the opposite side to the earth from the moon. The sun's effect on tides is about 50% of the moon's effect. Due to narrowness of the bays and river mouths (called estuaries), very high amplitudes of tides can occur.

The period of earth's rotation on its axis is about 23 hrs and 56 minutes while the moon's period of rotation round the earth is 24 hrs 50 minutes. Consequently, the high tides do not occur at the same time by the clock on all days, but are delayed by 54 minutes each day. Due to the intermittent nature of the power generation from a tidal scheme, storage of energy is required.

8.8 Geothermal Power

Geothermal energy comes from two sources: from radioactive decay in the crust of the earth, and from heat trickling through the mantle from the earth's core. The heat in the core arises from the fact that the earth used to be red-hot and is still cooling down and solidifying. The heat in the core is also topped up by tidal friction.

Geothermal energy is an attractive renewable because it is always on (unlike solar) and independent of weather. However, the constraint of speed at which heat travels through the solid rock limits the rate at which heat can be sustainably removed from the red hot interior of the earth. Typically, the heat flow from the centre coming through the mantle is about 10 mW/m^2. The heat flow at the surface of the earth is 50 mW/m^2. The radioactive decay has added 40 mW/m^2.

8.9 Energy Storage Systems

Both solar and wind power systems suffer from fluctuations in the power output due to variations in the sunlight and wind velocities. In the former, with PV panels, the output is zero during nights. In some places, it is observed that the wind velocities are higher in the nights thus compensating for the fall in the PV cell output. It is also possible that with a large number of wind turbines located at different locations may give a reasonably aggregate constant power.

The deployment of energy storage systems can compensate for the variation of power outputs of renewable power such as solar and wind. There are several types of electrical energy storage (EES) systems. A widely used method of classifying EES is based on the form of energy used. There are 5 classes of EES described below (IEC White Paper, 2011).

A. *Mechanical*: Includes Pumped Hydro (PHS), Compressed Air (CAES) and Flywheel (FES)

B. *Electrochemical*: Includes 1) Secondary batteries (Lead Acid(LA), Nickel Cadmium (NiCd) and Nickel Metal Hydride (NiMH) and 2) Flow batteries (Redox Flow and Hybrid Flow)

C. *Chemical:* Hydrogen Electrolysis, Fuel Cell and Synthetic Natural Gas (SNG)

D. *Electrical*: Includes Double Layer Capacitor (DLC) and Superconducting Magnetic Energy (SMES)

E. *Thermal*: Storage of sensible heat, storage of latent heat and thermo-chemical absorption

Pumped Hydro Storage (PHS)

The pumped hydro storage power plants represent nearly 99 % of world-wide installed EES capacity and which is about 3 % of global generation capacity. Conventional PHS use two water reservoirs at different elevations to pump water during off-peak hours from the lower to the upper reservoir (charging) and from the upper to the lower reservoir, powering a turbine with a generator to produce electricity (discharging). For lower reservoir, flooded mine shafts, other underground cavities and open sea are also technically feasible. The latter was first built in Japan in 1999. The first PHS plants were built in Italy and Switzerland in 1890s. Typical discharge times range from several hours to a few days. The efficiency of PHS plants is in the range of 70 to 85 %.

Compressed Air Energy Storage (CAES)

Compressed air energy storage (CAES) is a technology used since the 19th century for different industrial applications including mobile ones. Electricity is used to compress air and store it in either an underground structure or an above-ground system of vessels and pipes. When needed, the compressed air is mixed with natural gas, burned and expanded in a modified gas turbine. Typical underground storage options are caverns, aquifers or abandoned mines.If the heat released during compression is dissipated by cooling and not stored, the air must be reheated prior to expansion in the turbine. This process is called diabetic CAES and results in low round-trip efficiencies of less than 50 %. Diabetic technology is well proven; the plants have a high reliability and are capable of starting without extraneous power. In an adiabetic CAES system, currently under development, the released heat is retained in thermal storage and used again during expansion in a turbine. The advantage of CAES is its large capacity; disadvantages are low efficiency and geographical limitation of locations.

Flywheel Energy Storage (FES)

In flywheel energy storage (FES), rotational energy is stored in an accelerated rotor, a massive rotating cylinder. The energy is maintained in the flywheel by keeping the rotating body at a constant speed. To accelerate the flywheel, electricity is supplied by a transmission device; the energy is extracted by the same device by reducing the speed. Flywheels of the first generation (available since 1970), use a large rotating body on mechanical bearings. Advanced FES systems have rotors made of high-strength carbon filaments, suspended by magnetic bearings, and spinning at speeds from 20,000 to over 50,000 RPM in a vacuum enclosure. The main features of flywheels are excellent cycle stability and long life, little maintenance, high power density and the use of environmentally inert material. Flywheels are commercially deployed for power quality in industrial and UPS applications. Efforts are being made to optimize flywheels for long-duration operation (several hours) for use in vehicles and power plants.

Secondary Batteries

Lead acid (LA) batteries are the world's most widely used battery type and have been commercially deployed since 1890. They are used in both mobile and stationary applications. Their typical applications are: emergency power supply systems, stand-alone systems with Photo Voltaic (PV), mitigation of output fluctuations from wind power and as starter batteries in vehicles. Stationary lead acid batteries have typical service life of 6-15 years with a cycle life of 1500 cycles at 80 % depth of discharge. The cycle efficiency level is around 80 to 90 %. Lead acid batteries offer a mature and well researched technology at low cost. One disadvantage is that usable capacity decreases when there is a deep discharge. For example, if a battery is discharged in one hour, only about 50-70 % of the rated capacity is available. The other drawbacks are lower energy density and the use of lead, a hazardous material.

Nickel-Cadmium (NiCd) batteries had been in commercial use since 1915 before the commercial introduction of **Nickel Metal Hydride (NiMH)** batteries around 1995. Compared to lead-acid batteries, nickel-base batteries have a higher power density, a slightly greater energy density and higher number of cycles of use. However because of toxicity of Cadmium, Ni-Cd batteries have been prohibited in Europe since 2006. NiMH batteries were initially developed to replace NiCd batteries. They have all the good properties of NiCd batteries except for the maximum nominal capacity which is still ten times less compared to NiCd and LA batteries.

Lithium ion (Li-ion) batteries have become the most important storage technology in the areas of portable and mobile applications (lap top, cell phone, electric bicycle, electric car) since around 2000.High cell voltage levels up to 3.7 nominal voltage mean that one lithium ion battery can replace three NiCd or NiMH cells which has a cell voltage of only 1.2 V. Other advantages are their high gravimetric energy density and the cost reductions through mass production. They have a share of over 50 % in the small portable device market,there are still challenges for developing large-scale Li-ion batteries. The main obstacle is the high cost of over $ 600/kWh due to special packaging and internal overcharge protection circuits. Li-ion batteries have a high efficiency in the range of 95-98 %. Nearly any discharge time from seconds to weeks can be realized which makes them very flexible storage technology. However, safety is a serious issue in Li-ion batteries. Most of the metal oxide electrodes are thermally unstable and can decompose at elevated temperatures, releasing oxygen which can lead to a thermal runaway. Li-ion battery technology is still developing.

Metal air (Me-air) electrochemical cell consists of the anode made from pure metal and the cathode connected to an inexhaustible supply of air. For the electrochemical reaction, only oxygen in the air is used. Among the various metals, the lithium is most attractive as the theoretical specific energy of the Lithium -air battery is

11.14 kWh/kg which is about 100 times more than other battery types. It is even greater than that of petrol(10.15 kWh/kg).However, the high reactivity of lithium with air and humidity can cause fire. Currently only zinc-air battery with a theoretical specific energy of 1.35 kWh/kg is technically feasible. A satisfactory electrically rechargeable metal air battery has not yet been perfected for commercial use. It is to be noted that the oxygen does not contribute to the energy as it is not stored in the battery.

Sodium sulphur (NAS) batteries consist of liquid (molten) sulphur at the positive electrode and liquid(molten) sodium at the negative electrode; the active materials are separated by a solid beta alumina ceramic electrolyte. The battery temperature is kept between 300 and 350 degree C to keep the electrodes molten. These batteries have a life of about 4500 cycles and have a discharge time of 6 to 7.2 hours. The NaS battery technology has been demonstrated at around 200 sites in Japan, Germany, France and U.S. However, on September 21, 2011 there was fire involving NaS batteries manufactured by NGK, Japan (the sole manufacturing company). Since then NGK has decided to implement safety enhancement measures. Research is also in progress elsewhere to develop NaS batteries operating with lower temperatures.

Sodium nickel chloride (NaNiCl) battery, better known as ZEBRA (Zero Emission Battery Research) battery is similar to NaS battery-a high temperature battery and has been commercially available since 1995. Its operating temperature is around 270 degree C and it uses nickel chloride instead of sulpur for positive electrode.

Flow Batteries

In secondary batteries, the energy is stored and discharged in the active masses of the electrodes. In a flow battery, the energy is stored in one or more electroactive substances dissolved in liquid electrolytes which are stored externally in tanks and pumped through the electrochemical cell that converts chemical energy directly to electrical energy and vice versa. The power is defined by the size and design of the cell whereas the energy depends on the size of the tanks. Originally developed by NASA in the early 70s for long term space flights, flow batteries are now being considered for EES for durations of hours or days with a power rating of up to several MW. Flow batteries are classified into redox flow (RFB) and hybrid flow (HFB) batteries. A Vanadium RFB storage system of 500 kW and

10 hrs has been installed in Japan for power quality applications involving 3 MW, 1.5 s.

Chemical Energy Storage

Here, hydrogen (H_2) or synthetic natural gas (SNG) are used as secondary energy carriers. The "excess" electrical energy is used to produce hydrogen via water electrolysis. Once hydrogen is produced, different ways are available for using it as an energy carrier either as pure hydrogen or as SNG. Although the overall efficiency is low compared to storage technologies such as PHS and Li-ion, chemical storage allows storage of large amounts of energy ,up to the TWh range, and for greater periods of time-even as seasonal storage. Another advantage of hydrogen and SNG is that they can be used in other sectors such as transport, heating and chemical industry.

Hydrogen

A hydrogen storage system consists of an electrolyzer, a hydrogen storage tank and a fuel cell.

An electrolyzer is an electrochemical converter which splits water with the help of electricity into hydrogen and oxygen. It is an endothermal process, i.e. heat is required during the reaction. Hydrogen is stored under pressure in gas bottles and tanks, and this can be done practically for an unlimited time. To generate electricity, both gases flow into the fuel cell where an electrochemical reaction takes place: hydrogen and oxygen react producing water, heat is released and electricity is generated. For economic and practical reasons, oxygen is not stored but vented to the atmosphere on electrolysis, and oxygen from the air is taken for power generation. In addition to the fuel cells, gas motors, gas turbines and combined cycles of gas and steam turbines can be used for power generation. Hydrogen systems with fuel cells(less than 1 MW) and gas motors (under 10 MW) can be adopted for combined heat and power generation in decentralized installations. Gas and steam turbines with up to several hundred MW can be used as peaking power plants. The overall efficiency is around 40 %. For stationary applications, gaseous storage of hydrogen under high pressure is the most popular choice. Pressures up to 900 bars are used for storing hydrogen in small amounts. So far, no commercial projects are in existence for hydrogen storage systems.

Synthetic Natural Gas (SNG)

Synthesis of methane (also called SNG) is the second option to store electricity as chemical energy. Hydrogen produced by the electrolyzer is made to react with carbon dioxide to obtain methane in a methanation reactor. Just like hydrogen, SNG can be stored in pressure tanks underground or fed directly into the gas grid. Several CO_2 sources are possible for the methanation process such as fossil-fuelled power stations, industrial installations or bio-gas plants. The overall efficiency is less than 35 %.

Double Layer Capacitors (DLC)

DLC, also called as supercapacitors have been known for 60 years. They fill the gap between the capacitors used in electronics and general batteries because of their nearly unlimited cycle stability and extremely high power capability. They also have many orders of magnitude higher energy storage capability compared to traditional capacitors. They have capacitances of the order of several thousand farads, fast charge and discharge cycles due to very low inherent resistances. Other advantages are durability, high reliability, no maintenance, long life time and operation over a wide temperature range. The life time reaches one million cycles without any degradation. The efficiency is around 90 % and discharge times are in the range of seconds to hours. Their specific power density is about ten times higher than that of conventional batteries, but their specific energy density is about ten times lower. Since 1980, they have been applied in consumer and power electronics. A DLC is ideally suited as a UPS to bridge short term voltage failures. They can also be used in electric vehicles as a buffer for the acceleration process and regenerative braking.

Superconducting Magnetic Energy Storage (SMES)

The energy is stored in the magnetic field created by the flow of direct current in a superconducting coil, which is kept below the critical temperature (4 degree K for liquid helium). However, today high temperature superconducting materials are available now, that function at 100 degree K. Apart from the superconducting coil, power conditioning equipment and a cryogenically cooled refrigeration system is required in a SMES. The advantages of SMES are -fast response, high efficiency (85-90 %) and very high power output provided for short periods. The efficiency is affected by the energy demands of the refrigeration system. So far, small SMES products are commercially available for power quality control in microchip fabrication facilities.

Thermal Storage Systems

Thermal storage can be subdivided into different technologies: storage of sensible heat and storage of latent heat. The domestic hot water tank is an example of the former. The storage medium may be liquid such as water or thermo-oil, or solid such as concrete or the ground. Thermal energy is stored only through a change of the temperature of the storage medium. Latent heat storage is accomplished by using phase change materials (PCM) as storage media. There are organic (paraffins) and inorganic PCMs (salt hydrates) used for such systems. Latent heat is the energy exchanged during a phase change such as the melting of ice. Currently most PCMs use the solid-liquid phase change such as molten salts as a thermal storage medium for concentrated solar power (CSP) plants. CSP plants primarily produce heat, and this can be stored easily before conversion to electricity and thus provide dispatchable electrical energy. State-of-the art technology is a two tank system, with the molten salt as heat transfer fluid and storage medium. The molten salt is heated by solar radiation and the hot salt passes through a steam generator which powers a steam turbine. Subsequently, the cold salt (still molten) is stored in the second tank before it is pumped to the solar tower again.

Applications of EES

According to ERPI (USA) analysis, the applications of EES are classified under 3 heads as follows:

A. Generation and system level
B. T & D system applications
C. End-User applications

Different applications with different requirements demand different features from EES. Fig. 8.1 (reproduced from IEC White Paper, 2011, Fraunhofer Institute of Solar Energy, Germany) gives a comparison of power, energy content and discharge time of different technologies. This is a double logarithmic chart where rated power is plotted against the energy content of electrical energy storage systems. The nominal discharge time at rated power is also shown, covering a range of seconds to months. Fig. 8.1 gives not only application areas of present EES systems, but also the predicted range in future applications. Not all technologies are commercially available in the ranges shown in the figure, at present. However, they are expected to play a major role in the future.

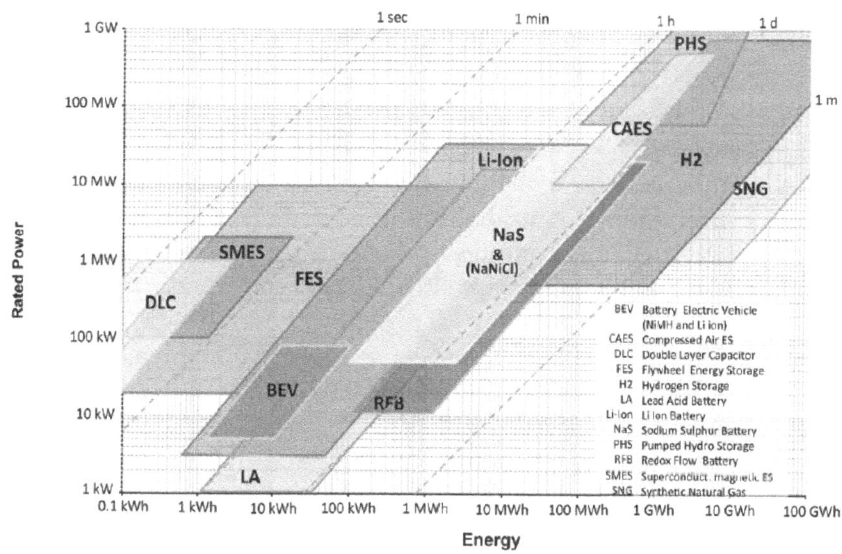

Fig. 8.1 Comparison of rated power, energy content and discharge times

According to the report by International Electrotechnical Commission (IEC), the major objectives of EES are

1. Carbon dioxide reduction
2. More efficient and reliable energy

The drivers for developing the EES technology are (a) renewable power (b) smart grid and (c) dispersed generation/micro-grid. Many studies have shown that EES is indispensible for the introduction of large amounts of renewable energy. Therefore the necessary volume and timing of EES is strongly dependant on the pace of renewable energy development.

To conclude, the EES technologies (available or envisaged) can be categorized into 3 categories based on the discharge time.

1. *Short discharge time (seconds to minutes)*: double-layer capacitors (DLC), superconducting magnetic energy storage (SMES) and flywheel (FES). The energy to power ratio is less than 1.
2. *Medium discharge time (minutes to hours)*: FES and for large capacities -electrochemical EES which is the dominant technology: Lead-Acid (LA), Lithium Ion (Li-ion) and sodium sulphur (NaS) batteries. Average energy to power ratio of between 1 to 10.
3. *Long discharge time (days to months)*: hydrogen (H_2) and synthetic natural gas (SNG). For these systems, the energy to power ratio is considerably greater than 10. Pumped hydro storage (PHS), compressed air energy storage (CAES) and redox flow batteries are situated between storage systems for the medium and long discharge times. Like H_2 and SNG systems, these EES systems have external storage tanks. But the energy densities are rather low which limits the energy to power ratio between 5 and 30.

Discussion

According to a white paper by EPRI, U.S. (December, 2010) the worldwide installed capacities of prominent EES are:

PHS- 127,000 MW, CAES- 440 MW, NaS Battery- 316 MW, Lead-Acid Battery-35 MW, NiCd Battery- 27 MW, FES- < 25 MW, Li-ion Battery – 20 MW, Redox Flow Battery - < 3 MW.

This shows that over 99 % of the EES is PHS, based on mechanical storage of energy. The flow batteries and chemical energy storage (Hydrogen and SNG) are very promising, but still lot of research and development is required before they can be applied.

There are some basic issues in the application of EES in power distribution systems. The first is the cost factor as it is involved in the techno-economic analysis. A recent publication by Oberschmidt et al (2013) describes the application of EES in power trading and mention three applications: (i) by wind energy operators to level out deviations from planned output capacity due to forecasting errors, (ii) by large industrial consumers to avoid peak overloads and thus avoid high energy prices charged for peak overloads and (iii) by distribution system operators to guarantee uninterruptible power supply and to optimize the operation.

There are environmental aspects of EES depending on the type of technology used. The manufacture, use and disposal of EES systems can cause environmental effects as well as health and safety hazards. The electricity storage always results in a loss of energy which depends on the round-trip efficiency. The resource requirements for electrochemical energy storage are critical in terms of vulnerability and potential hazards. The electrodes of batteries are usually made of metals or metal oxides, which are placed in an electrolytic solution. The requirement of Vanadium based redox flow batteries (RFB) exceeds the present global production of Vanadium.

The health and safety issues also depend on the type of EES. For example, Lithium ion systems are generally less robust compared to other battery types due to high energy densities causing higher accident hazards. The lead used in LA battery is hazardous. Sodium Sulphur (NaS) batteries contain flammable materials and operate at high temperatures as mentioned earlier.

Finally, it needs to be said that the costs have to be reduced significantly for EES, in order to provide competitive solutions. A comprehensive analysis of the whole life cycle as well as specific operating conditions must be considered before choosing a specific technology.

CHAPTER 9

Restructuring and Competition

"My mind is made up. Don't confuse me with facts"

Anon

9.1 Introduction

Electric utility restructuring was initiated in the 1990s, ostensibly to solve the problem of high electricity costs. In U.S.A, the most affected states are in the Northeast and California. By treating electrical energy as a product to be sold in the market place where the prices are determined by the balance between demand and supply, it was felt that the prices will fall and the customers will benefit if the utilities are restructured by allowing competition. However, the electrical energy is unlike any other product where the customers can benefit by free market. In the first place, there is no distinction between a unit of energy produced by renewable power (called green) and that produced by a plant using coal as fuel (called brown power). The transmission network that comes between the generating plants and the customer loads introduces further complication in the sense that a user at a specific location cannot find out the source(s) of the received energy. The non-availability of viable energy storage technologies and the requirement of synchronous operation in a AC grid further complicate the problem. So it is not surprising that contrary to the hopes of politicians, high cost states have seen little price relief and competition had a negligible impact on prices. As a matter of fact, the restructuring contributed to the severity of the 2000-2001 California electricity crisis and the August 2003 blackout in the Northeast. Before restructuring, the electricity markets were governed by regulated natural monopoly. The electric utilities were vertically integrated utilities (VIU) that owned generation, transmission, distribution and sale of energy in a specified area. The utilities may be owned privately or by public (including government). There are 3 models of restructuring the electricity market.

1. Single buyer, competition is only among generators (energy producers)
2. Competition in the wholesale market
3. Competition in the wholesale and retail market

As discussed in Chapter 1, the electricity supply industry (ESI) is governed by the following features:

1. The power demand at each location in the system network has to be met practically instantaneously.
2. The generators have to operate with adequate spinning reserves to take care of the emergencies (also called as contingencies).
3. The system has to operate stably under normal and emergency conditions. There are three aspects of stability - (a) maintaining synchronous operation (b) preventing voltage collapse and (c) controlling frequency within operating limits.

A major complication in operating the power systems arises due to the role of electricity in economy and society. The damage caused by power interruptions is many times larger than the cost of electric energy not supplied.

9.2 Reasons for Restructuring and Deregulation of ESI (US Experience)

According to the paper entitled 'Rethinking Electricity Restructuring' by Peter Vom Doren and Jerry Taylor published in Policy Analysis (Cato Institute) in November 30, 2004, "Electric utility restructuring was a political answer of high rates in the Northeast and California. Firms threatened to leave high cost states, so those states attempted to bring low-cost electricity to the firms. Under restructuring, local electricity generators would no longer have a monopoly over local customers. In theory, distant (lower-cost) generators could compete for business and rates would go down".

Economists argued that monopolistic structure of ESI encouraged excessive investment in generation as the investment in capital received a guaranteed return. The energy crisis precipitated in 1973 by high increase in the imported oil (from $2 to $20 per barrel) led to a flurry in U.S utilities to build coal based and nuclear thermal power plants. However, the growing hostility to coal fired generation for environmental reasons and nuclear plants due to safety issues and high costs, forced some utilities to junk nuclear power. There was resistance to the demand of stranded costs when legislators forced shut down of nuclear plants.

Economists also argued that prices for electricity did not serve their usual role of signaling to the consumers the marginal costs of additional consumption which vary by time of the day and season. It was argued that the fixed rates were too low on peak and too high off peak. It was hoped that market forces would introduce marginal-cost pricing and as a result reduce peak demand and increase off-peak demand.

The deregulation of interstate wholesale electricity markets in 1992 and the restructuring of state-level regulation, where it has occurred, have induced owners of generators to think about costs and risk. In the 1992-2002 period, investors added too much capacity resulting in low wholesale prices and widespread bankruptcy in the electric generation sector.

9.3 Outcomes of Restructuring in U.S.A

Peter Van Doren and Jerry Taylor claim that restructuring has delivered some of its promised benefits. But they also observe that "Most Americans associate markets in electricity with bad outcomes because of California and the major 2003 Northeast blackout. Accordingly, anyone who believes market forces ought to play a larger role in electricity has to argue convincingly that

1. The California meltdown and the Northeast blackout were not result of market forces.
2. The low costs of states still under the old regulatory regime are not the result of regulation.
3. Gains to trade (efficiency improvements) not possible in the regulated status quo, would take place in a truly deregulated world; and

4. The "commons" nature of the AC transmission system can be managed in a deregulated world."

The authors state that this is not easy. They also admit that "Vertical integration may be the most efficient organizational structure for the electrical industry". They also state "The differences between a regulated and an unregulated market may not be as great in the electricity sector as they are in other markets".

In a paper entitled "What's wrong with electricity grid" (Industrial Physicist, American Institute of Physics, Oct/Nov. 2003), Eric Lerner states "Experts widely agree that such failures (as on 14th August 2003), of the power transmission system are a nearly unavoidable products of a collision between the physics of the system and economic rules that now regulate it. To avoid future incidents, the nation must either physically transform the system to accommodate the new rules, or change the rules to better mesh with the power grid's physical behavior". The author also states "After wholesale electricity trading began in earnest following Federal Energy Regulating Commission's Order 888, stress on the transmission grid jumped and continued to climb (as shown by the transmission loading relief procedures) and also, monthly average frequency errors.

9.4 Indian Experience

Restructuring of ESI in U.K and India was mainly motivated by the need to privatize the sector to improve efficiency or introduce competition by bringing in private capital as the government cannot meet the requirements of fast growth of installed generation capacity necessitated by the growth in economy.

In India two issues, one related to the transmission and distribution (T & D) power losses and other concerning plant load factor (PLF) have bothered the decision makers responsible for utilities owned by the state or central governments. There were incentives offered to plant operators to maintain high PLF. This resulted in wide fluctuations in the frequency as there are no load frequency controllers. The turbine governors were prevented from free operation that would reduce power output of the generator if the frequency went above the rated value of 50 Hz. There are 5 regional electricity boards in India supplying - North, South, West, East and North-East regions.

The Eastern region had higher generation compared to the load and often the frequency could rise to 53 Hz as the generation was not backed down. The national power utilities - NTPC, NHPC and NPC supplied power to state utilities which were perennially short of generation due to inadequate capital investment. Free supply of power for agriculture which was not metered and other commercial losses meant that states were heavily dependent on central power generation. In AC power systems, it is easy for a state utility to draw power above the scheduled interchange. The unscheduled interchange (UI) also contributed to wide frequency fluctuations and inefficient utilization of electric energy in the overall system. There were frequent blackouts.

9.5 Availability Based Tariff (ABT)

To improve the system operation by discouraging UI and also ensure that PLF is controlled in response to the load demand, the availability based tariff (ABT) was recommended by M/s. ELC of U.S.A in 1994 and it was implemented from 1st July 2002 by Powergrid company .The ABT charges UI according to a formula which ensures frequency regulation. The rate for UI s zero if the frequency exceeds 50.5 Hz and increases as the frequency is reduced. There is a ceiling 49.0 Hz (which represents the lower limit on frequency). The rates and limits have been modified over the years.

Introducing competition by allowing independent power producers (IPP) or captive power plants to sell their power to state utilities and allowing access to the transmission facilities created by the state or central government, allowed the ABT to function effectively. The electricity act of 2003 permitted power exchange for power trading and the private entities can generate and sell power in addition to build transmission lines where necessary. The Act enabled separation of generation, transmission and distribution by breaking up existing state electricity boards which were working earlier as vertically integrated utilities (VIU).

9.6 A Comparison of Monopolistic and Competitive Markets

Before outlining the differences between the two types of markets, let us look at the similarities. These are:

1. If the objective is to benefit consumers, they want reliable power supply with minimum or affordable costs.
2. The costs in ESI are approximately 50-60% for generation, 40-45% for transmission and distribution and 5% for sale of electric energy.
3. The state regulation is required in both types of markets. As a matter of fact, Casazza et al (2007) state that "The word 'deregulation' is a misnomer. Under its guise, the federal government has taken over effective control of the electric power industry. Never in the history has the U.S government so regulated power supply. Thousands of pages of new regulations have been promulgated by Federal Energy Regulatory Commission (FERC) and thousands of pages of rules and procedures have been imposed by NERC (North American Electric Reliability Council). NERC's Energy Regulatory Office (ERO) application alone runs to over 10,000 pages. It is not surprising that proponents nowadays rarely use the term 'deregulation', preferring the seemingly more benign 'restructuring' instead".

In principle, both producers and consumers of energy should benefit from restructuring leading to competition. The competitive markets are advantageous if the benefits exceed the expenditures on market infrastructure and the negative consequences due to imperfection in the market design (gaming, for example).

Advantages of VIU (Regulated monopoly)

1. Ability to gainfully employ power plants of different types of plants to meet base, intermediate and peak loads. This benefits both the utility and customer as the customer has to bear only average costs of generation rather than pay marginal costs of the least efficient power plant.

2. Ability to optimize the system operation by long term planning of addition of new power plants and new transmission lines for efficient delivery to meet the load growth in industrial, commercial and residential sectors.
3. There are no disadvantages which are associated with the creation of a competitive wholesale market (Model 2), such as increased production costs due to competition among buyers. In India, for example, the spot market prices have touched over Rs. 10-13 per unit. Some private companies have recovered their capital costs within 2 years by just supplying energy through power exchanges (without entering into bilateral contracts that would assure them of a steady power demand).
4. VIUs can exploit economies of scale wherever possible.
5. Competitive markets can result in worsening of reliability. Power blackouts were observed in USA, Canada and Western Europe in 2003, Russia in 2005.

Belyaev (2007) has shown that single buyer market (Model 1) only can provide the benefits to consumers compared to Models 2 and 3. As a matter of fact, Model 3, which involves competition in retail market, is not realistic.

Present Status on Restructuring

France and Japan in addition to many states within USA still operate VIUs. Many countries have restrained the restructuring process to only model 1 (single buyer market), for example, South Korea and China. Countries in S. America - Chile, Argentina and Brazil have experienced capacity and energy shortages and the governments have restored price regulation.

Indian Experience

After the passing of Electricity Act in 2003, the power sector has been restructured with separation of state utilities into generation, transmission and distribution companies. Although most of the DISCOMs (Distribution Companies) are owned by the state governments, some DISCOMs are privately owned, e.g. in Mumbai and Delhi. The private companies can set up generation and transmission facilities and sell power through power exchanges (similar to stock exchange) to the highest bidders. The total capacity has reached nearly 2,30,000 MW (by end of September 2013) that includes nearly 73,000 MW of private generation. The renewable energy source (RES) component is about 28,184 MW including 24,382 MW by private investment. The frequency variation is significantly reduced and the average frequency is nearly 50 Hz.

Yet, the reliability of electricity supply has not improved. There are perennial shortages in practically all the states. The system operation is beset with several problems. A major concern is the power loss which includes technical (I^2R) loss and commercial loss (euphemism for theft). The government has introduced 'Restructured Accelerated Power Development and Reforms Program (RAPDRP)' to improve the performance of power distribution sector and reduce power losses. The cost benefit analysis of the investments is not published.

There were two major power blackouts on July 30 and 31, 2012 involving up to 650 million customers in the 3 regions - North, East and Northeast. (These regions are synchronized along with Western Region). Only southern regional grid is connected to Western and Eastern grids by 3 HVDC links (2 back to back and one long distance link). There are plans to also synchronize the southern region into a national grid.

9.7 Decline in Reliability of Bulk Power Systems

Casazza et al (2007) state "Deregulation and the concomitant restructuring of the electric power industry in the United States have resulted in a decline in the reliability of North American bulk power systems and constitute the ultimate root cause of the California meltdown, Enron's depredations and the 14th August 2003 blackout".

They cite the following reasons for the problem:

1. Corporate culture in the electric power industry has shifted from coordination and cooperation to competition and confrontation.
2. Industry expenditure for new facilities and manpower has been reduced.
3. The qualifications of senior managers have shifted away from a working knowledge of the technical aspects of power systems, and towards marketing and legal issues.

4. Institutional knowledge, including lessons learned from past blackouts, has been lost.
5. The increased number of players and extensive new regulations have vastly increased the complexity of both decision making and day-to-day operations.
6. North American Electric Reliability Council (NERC) has already watered down its reliability standards and some wish to reduce it even further.

9.8 Recent Developments

Following major blackouts in many parts of the world, the decision makers (primarily politicians supported by lawyers, investment bankers, economists, etc) are putting the blame on electrical (power) engineers for the debacle of restructuring and pushing for ICT (Information and Communication Technology) to reform the grid so that it becomes 'smart' using 'technology'. However, it is not obvious what the plan of action is or the costs involved in improving the grid. While it is well known that high power electronic controllers (HVDC and FACTS) can improve system performance, more such controllers have been installed outside U.S.A (for example, China and India) than within. Appendix D describes an interesting thought experiment. (DOE U.S. - Introduction to Smart Grid) that compares the states of communication technology with power engineering technology. It is amazing that in this information age, such ignorance exists even among educated people (see also the postscript to Appendix D).

Unfortunately, the American Public appear to be ignorant about the various technology missions undertaken in the post Second World War period (some of which have yet to be applied as reliability and costs are major considerations that inhibit). The nuclear power based on fission is a major development that has provided significant amounts of base load (20-30%). In France, 75% of power is met by nuclear power plants. The application of superconductivity for synchronous generators, magnetic energy storage and underground cable transmission has been investigated extensively in the seventies. The development of high power electronic controllers since early seventies for HVDC transmission and FACTS controllers is still continuing and has been applied. Magnetic levitation and linear motors have been developed for high speed vehicles in electric traction. The energy crisis in early seventies led to major developments in renewable energy. The only hitch has been in the field of energy extraction from nuclear fusion. However, research is still continuing on the application of Tokomaks.

The application of computers for improving the system operation and enhance security, started after the major blackout in 1965 in North East U.S.A. There is continuous progress in developing analytics incorporating new high power electronic controllers such as HVDC and FACTS for increased and secure power transfer from generating stations. The work on Distribution Automation (DAC) has been in progress since early eighties. The communication technology has been applied for a long time for system protection and control. However, the introduction of a new technology in a big way in electricity supply systems, has to pass the test of reliability and cost-benefit analysis. Thus, while it may be feasible for manufacturers of cellular phones to introduce newer models every year, the communication systems in electricity supply have to be chosen with sufficient care to ensure the technology does not get obsolete soon. The key word is 'evolution' rather than 'revolution'. The formulation of international standards and 'interoperability' is a must before a major overhaul involving massive investments in a new technology.

The main reasons for ignoring these developments appears to be two fold: 1. the electricity supply industry (ESI) is not attracting the best talent as research in the power area has been neglected in universities and 2. failure of the ESI management to present effectively the problems affecting the growth of demand and a plan of action to tackle these problems. The citizens question why deregulation/restructuring that has been implemented successfully in the communication industry, cannot be achieved in ESI. A major difference, for example, is that generation of a signal requires practically no energy, but can require substantial energy for signal transmission through communication media (using fiber-optic cables or wireless). As discussed in chapter 6, the switches/relays used in telephony, can operate continuously as very low voltages and currents are involved; however, the circuit breakers (CB) dealing with large currents(thousands of amperes) are different. Power semiconductor switches have severe limitations at present and cannot be expected to replace mechanically operated CBs in high voltage networks. Interestingly, most of the present research is on nano-materials and very little funding is available for power semiconductors that could perhaps replace mechanical CBs in medium voltage networks.

9.9 Conclusions and a Road Map for the Future

The changes introduced by the politicians (supported by economists and lawyers) in the nineties for restructuring of ESI, are here to stay and is governed by the concepts of free markets and individual choices driving the economic activity and conduct of public enterprises. While the United States of America has been at the forefront of free enterprise and unhindered trade in all economic activity, it has also permitted regulated monopoly in electricity supply and telecommunications since the beginning. The technological innovations have permitted the restructuring of telecom industry and this has given the impetus to extend it to the ESI. However, the two are not comparable, not because of lack of technological innovations, but cost considerations that discourage massive investments in the electricity supply sector. The smart grid initiatives have been supported by the U.S. federal grants in 2009 as a part of the stimulus package to recover from the economic collapse. The renewables (such as wind and solar) are not capable of providing firm (base load) power and hence cannot replace in the near future coal or gas based power plants. The new electrical energy storage (EES) technologies are not yet mature and cost effective. Hence, nuclear power is the only solution for providing base load power without increasing carbon emission.

The smart grid solutions are not necessarily novel in concept as demand side management (DSM) has been practiced for a long time with analog signals to switch on/off heating and cooling loads. Apart from the use of digital signals and the ICT, the only change relates to the possibility of the customers connected at low voltages also supplying energy to the microgrid at times when the local generation(such as roof top solar photo-voltaics) exceeds the consumption. However, it may be more appropriate to store the excess energy during periods of low consumption. The role of the so called "smart meters" is not well defined. It is essential to define the functionalities of the communication channel between the utility (DISCOM) and the individual customers so that cost effective measures can be adopted. For example, in many places including India, a customer may need only one-way communication from the utility about the supply constraints and whether the customer needs to cut down on his energy demand. It is to be noted that "restructuring" implies the customers must provide for alternate supply (locally) when the power utility is unable to meet the demand. In other words, the utility will not be responsible for power failures and they can charge more for better reliability. In the past, there was a proposal to charge more for improved power quality (PQ) and provide computer grade power as the utility had to invest in FACTS controllers and other devices to be located in distribution systems to improve PQ (see article by Hingorani (1995), IEEE Spectrum). Most of the improvements to be provided by smart grids relate to the Distribution Automation which was discussed in early eighties.

Location of Transmission Lines

A major bottleneck in power delivery is the long times involved in approval of the rights of way in the construction of new transmission lines. Instead of just building a new line from a generating station to a load centre, it makes more sense to plan a grid of HV or EHV AC lines just as one plans national highways. Obviously, this requires the government to play a major role in planning and executing the projects for bulk power transfer from several generating stations to the load centres from where it can be distributed using medium voltage feeders. The generation companies should be encouraged to locate the generating plants close to a transmission line. The power from a generating plant can be fed to a switching station through short HV lines. The capacity of a transmission corridor can be raised by a) building parallel lines wherever feasible or b) raising the voltage levels and/or conversion of AC to DC line where feasible, when ROW is a problem. It would be also worthwhile to examine the feasibility of installing power cables under the towers along the same ROW.

Finally, there is no short cut to practice energy conservation and limit the needs for energy in day to day living.

APPENDIX A

Fundamental and Secondary Units

There are basically six fundamental quantities: length, mass, time, electric current, temperature and luminous intensity. The basic units for these quantities are expressed in International Systems of Units (SI), which is a modernized version of metric system. The abbreviation SI is from the French name Systeme Internationale.

A.1 Fundamental Units

Length (L): Meter (m) = $10^2 cm$ = $10^3 mm$

Mass (M): Kilogram (kg) = 10^3 *gram*

Time (T): Second (s)

Current (I): Ampere (A)

Temperature: Kelvin (K)

Luminous intensity: Candela (cd)

The electrical and magnetic quantities can be expressed in terms of 4 units (meter-kilogram-second-ampere) (mksa), a sub system of the SI.

Metric Prefixes

For convenience metric prefixes are used wherever convenient. Table A.1 shows the prefixes, their magnitudes and abbreviation.

Table A.1 Metric Prefixes

Prefix	Magnitude	Abbreviation
exa	10^{18}	E
peta	10^{15}	P
tera	10^{12}	T
giga	10^{9}	G
mega	10^{6}	M
kilo	10^{3}	k
milli	10^{-3}	m
micro	10^{-6}	μ
nano	10^{-9}	n
pico	10^{-12}	p
femto	10^{-15}	f
atto	10^{-18}	a

A.2 Secondary Units

Mechanical Units

(See Table A.2)

Table A.2 Mechanical Units

Quantity	Description	SI unit	Equivalent unit
Acceleration	$\frac{\text{length}}{\text{time}^2}$	meter/sec^2	$\frac{N}{kg}$
Force	mass × acceleration	newton (N)	10^5 *dyne*
Energy	force × length	joule (J)	10^7 *ergs*
Velocity	$\frac{\text{length}}{\text{time}}$	meter/sec	
Power	$\frac{\text{energy}}{\text{time}}$	watt (W)	Nm/sec
Frequency	$\frac{\text{cycles}}{\text{time}}$	hertz (Hz)	cycles/sec
Period	$\frac{1}{\text{frequency}}$	second (s)	

Electrical Units

(See Table A.3)

Table A.3 Electrical Units

Quantity	Description	SI unit	Equivalent unit
Charge	currnet×time	Coulomb (C)	6.25×10^{8} electron charges
Potential	$\frac{\text{energy}}{\text{charge}}$	Volt (V)	$\frac{W}{A}$
Impedance	$\frac{\text{Potential}}{\text{Current}}$	Ohm (Ω)	$\frac{V}{A}$
Admittance	$\frac{1}{\text{impedance}}$	Mho (℧)	$\frac{A}{V}$
Capacitance	$\frac{\text{charge}}{\text{potential}}$	Farad (F)	
Resistance	$\frac{\text{Potential}}{\text{Current}}$	Ohm (Ω)	$\frac{V}{A}$
Conductance	$\frac{1}{\text{Resistance}}$	Mho (℧)	$\frac{A}{V}$
Reactance	$\frac{\text{Potential}}{\text{Current}}$	Ohm (Ω)	$\frac{V}{I}$
Suceptance	$\frac{\text{current}}{\text{voltage}}$	Mho (℧)	$\frac{I}{V}$

Magnetic Units

(See Table A.4)

Table A.4: Magnetic Units

Quantity	Description	SI unit	Equivalent unit
Magnetic Flux		Weber (Wb)	$\frac{Nm}{A}$
Flux Density	$\frac{\textit{Magnetic}\ \text{Flux}}{Area}$	Tesla (T)	$\frac{N}{Am}$
Flux linkage	$Flux \times turns$	Weber turns (Wb turns)	
Magnetic Field	$\frac{MMF}{length}$	$\frac{\text{Ampere}}{Meter}$	$\frac{A}{m}$
Inductance	$\frac{\textit{Flux}\ \text{linkage}}{Current}$	Henry (H)	$\frac{Wb}{A}$
MMF		ampere turn (A turn)	
Reluctance	$\frac{MMF}{Flux}$	$\frac{1}{Henry}$	$\frac{A}{Wb}$
Permeance	$\frac{Flux}{MMF}$	Henry (H)	$\frac{Wb}{A}$
Permeability	$\frac{\text{Inductance}}{length}$	Henry /meter	$\frac{Wb}{Am}$

APPENDIX B

Electric Shocks and Safety

B.1 Introduction

The tremendous increase in the use of household electric appliances, the proliferation of current operated tools in industry, and the continual introduction of new techniques for hospital care have created an additional hazard to human life.

The way people respond to electrical stimulation depends on several factors, including the manner in which the stimulation is introduced. The threshold of human reaction can be classified into four major categories according to how the stimulation is applied:

1. Contact currents enter a person through electrical conductors contacting the skin.
2. Capacitive spark discharges introduce transient currents into the body through an arc gap when the electrical breakdown potential of air is exceeded.
3. Electric fields on the surface of the body exert forces on hair and induce currents.
4. Magnetic fields induce internal body currents.

The response to direct contact is important in establishing safety limits on the direct effects of electric currents. Spark discharges may be of greater importance in establishing thresholds of annoyance or uncomfortable sensations or secondary effects due to startle reactions. For example, a workman may fall from a ladder. The direct effects of electric and magnetic fields are of minor importance for short-term stimulation at levels produced within power transmission corridor.

B.2 Human Response to Stimulation by Power Frequency Currents

The qualitative definitions of human response to stimulation by power frequency (50 or 60 Hz) currents are given below (in the order of increasing stimulation).

1. *Perception:* The person is just able to consciously detect the stimulus at a given detection probability. The sensation at a perception threshold is gentle and not necessarily unpleasant. It establishes a stimulation level below which there are no important short term effects.
2. *Annoyance:* The person would consider the sensation to be a mild irritant if it were to occur repeatedly.
3. *Startle:* If the stimulus occurred unexpectedly, it would probably produce an unintentional muscular reflex that could be hazardous under a defined set of circumstances.
4. *Aversion:* If a person receives one exposure, he would be motivated to avoid situations that he felt would lead to a similar experience.
5. *Let-go:* A person cannot let go of a gripped conductor as long as the stimulus persists, because of uncontrollable muscle contraction.
6. *Respiratory Tetanus:* A person is unable to breathe as long as the stimulus is applied, because of contraction of the muscles responsible for breathing.
7. *Fibrillation:* Uncoordinated asynchronous heart contractions that produce no pumping action.

The categories 3, 5, 6 and 7 relate to involuntary reactions. It is to be noted that the response of 'aversion' includes experience that one would describe as painful. The converse is not necessarily true. It is possible that an aversive electrical stimulus does not necessarily excite the unique receptors associated with the sensory experience of pain.

Factors Affecting Severity

When the body becomes a part of the electric circuit, the effects of electric current are largely due to the magnitude of the current and duration of the shock. The current depends on the voltage applied (which is generally known) and the impedance of the total path (of current) including the human body. The impedance of the human body is the major factor as the impedance of the power circuit is usually negligible. At power frequencies and voltages of 120-240 volts, the resistance at contact locations is the major factor limiting the current. Moist or liquid contacts such as those commonly experienced in bathroom, kitchen or garden constitute a potentially hazardous situation for receiving an electric shock. In contrast, the resistance at contact locations on both high-voltage and high-frequency circuits is relatively unimportant. At 240 volts and above, the voltage punctures the skin instantly, often leaving a deep localized burn. Here, the internal impedance of the body is the major current limiting factor. On currents of power frequency, the impedance of the body is essentially resistive, but above 1000 Hz, the body, because of its cellular structure, begins to exhibit a nonlinear characteristic. It has been shown that the decrease in impedance may be more than 50% for an increase in frequency from 50 Hz to 50,000 Hz.

For dry, intact skin, the resistance through limbs (hand to hand or hand to foot) may exceed 100,000 ohms per square centimeter. However, sweaty or broken skin may lose much of its insulating property and the electrical resistance of the underlying flesh may be more important. A value of 1000 ohms is representative of inter-limb resistance of a sweating or wet person; however, in evaluating electrical safety, an inter-limb resistance of 500 ohms is often used to represent the broken skin. The skin may be rapidly damaged by blistering or burning, and the value of 500 ohms is frequently used in estimating shock currents during industrial accidents.

The important parameter in defining most reactions to direct contact is the current density within the body. This may depend on the current level, the part of the body through which is introduced, the surface area and the geometry of the electrical contact.

Table B1 summarizes threshold reactions that are broadly classified as perception, startle, let-go, respiratory tetanus and fibrillation.

Effects of Higher Currents

Currents considerably in excess of those just necessary to produce ventricular fibrillation may cause cardiac arrest, respiratory inhibition, irreversible damage to nervous system, serious burns and unconsciousness. However, no numerical data is available.

Effects of Frequency

Studies published in 1971 by Geddes and Baker at Baylor College of Medicine in Texas show that the current required to produce fibrillation in dogs at 3000 Hz is 22-28 times that at 60 Hz depending on contact locations.

B.3 Reaction to Spark Discharges

When a person is within or near the right-of-way of transmission lines, capacitive spark discharges may occur between the person and a conducting object. In these cases, the person and/or the object act as capacitors on which a charge is induced by the action of the electric field produced by the transmission lines. When the person and the conducting object come close together (fraction of a centimeter), the electric breakdown potential of air may be exceeded, and an electric arc results. These discharges are qualitatively similar to the static discharges that occur when a person walks across a carpet on a dry day, except that a carpet spark is a single transient event, whereas an arc spark discharge from transmission lines is continued as long as a small air gap is maintained.

The threshold of human reaction to spark discharges will be different for AC and DC capacitive discharges. A DC discharge is a single transient event, whereas the AC discharges that are induced by high voltage transmission lines may occur every one-half cycle of the alternating voltage.

B.4 Reduction of Shock Hazards

At present, eight means of reducing the hazard of electric shock are: 1. isolation 2. guarding 3. insulation 4. grounding 5. double insulation 6. shock limitation 7. isolation transformers and 8. high frequency/direct current. The first four are of primary importance in controlling the hazards due to high voltage. Double insulation is largely applicable to low voltage hand tools and appliances.

Shock limitation can be achieved by the ground fault interrupter (GFI) in the USA and ELCB (Earth Leakage Circuit Breaker) in Europe and elsewhere. It is a fast-acting circuit breaker activated by currents of only milliamperes in magnitude flowing to ground.

Table B1: Threshold levels for 60 Hz Contact currents

Reaction/ Sensation	rms Current (mA)	Threshold Reaction/Sensation
Perception	0.09	Touch perception for 1% of women
	0.13	Touch perception for 1% of men
	0.24	Touch perception for 50% of women
	0.33	Grip perception for 1% of women
	0.36	Touch perception for 50% of men
	0.49	Grip perception for 1% of men
	0.73	Grip perception for 50% of women
	1.10	Grip perception for 50% of men
Startle	2.2	Estimated borderline hazardous reaction, 50% probability for women (arm contact)
	3.2	Estimated borderline hazardous reaction, 50% probability for women (pinched contacts)
Let-Go	4.5	Estimated let-go for 0.5% of children
	6.0	Let-go for 0.5% of women
	9.0	Let-go for 0.5% of men
	10.5	Let-go for 50% of women
	16.0	Let-go for 50% of men
Respiratory Tetanus	15	Breathing difficult for 50% of women
	23	Breathing difficult for 50% of men
Fibrillation	35	Estimated 3-s fibrillating current for 0.5% of 20-kg (44-lb) children
	100	Estimated 3-s fibrillating current for 0.5% of 70-kg (150-lb) adults
Established Standards	0.50	ANSI standard for maximum leakage (portable appliances)
	0.75	ANSI standard for maximum leakage (installed appliances)
	5.0	NESC recommended limit for induced current under transmission lines.

APPENDIX C

Nuclear Energy

C.1 Elements of Atomic Physics

Rutherford proposed in 1911 an atomic model, according to which, the atom consists of a heavy nucleus of radius in the order of 10^{-15}m (1 Fermi or Femtometer (fm)). The density of nuclei is nearly constant (around 10^{17} kg/m^3). In contrast, water has a density of 10^3 kg/m^3. The nucleus consists of positively charged protons (Z in number) and electrically neutral neutrons. The nucleus is surrounded by Z electrons and then the atom is electrically neutral on the whole. Since the nucleus consists of only positively charged protons, there must be only repulsive electrostatic force between the protons. However, since the nucleus is held together inside a minute space, there must be a very strong binding force to counteract the electrostatic force of repulsion. This is called the strong interaction (a new kind of force in the nature). The number of protons is called as the atomic number, N+Z (where N is the number of neutrons) is called the atomic mass. Isotopes of an element have the same atomic number, but have different atomic masses due to different number of neutrons in the nucleus.

C.2 Nuclear Reactor

The nuclear energy can be clubbed into two categories: (a) nuclear fission where uranium, an exceptionally heavy element is used as fuel, and (b) nuclear fusion where light elements, especially hydrogen would be used. The nuclear reactors built so far for the production of electric energy use fission. The prototype fusion reactors are yet to be built.

Nuclear power, fission or fusion has an important property, namely, the nuclear energy available per atom is roughly one million times bigger than the chemical energy per atom of typical fuels. This implies that the amounts of fuel and waste that must be dealt with a nuclear reactor can be one million times smaller compared to amounts of fuel and waste at a fossil-fuel power station of equal rating.

Fission Reaction

Naturally occurring uranium (whose atomic number is 92) contains 3 isotopes, U^{234}, U^{235} and U^{238}. (The superscript indicates atomic mass). However, the percentage of U^{238} is highest (99.3%). Of the three isotopes, U^{235} undergoes spontaneous fission when subjected to bombardment by slow neutrons. U^{235} is the naturally occurring fissile material. The break up of U^{235} yields fission products, neutrons and the release of a large amount of energy as heat (8.2×10^7 kilo Joules per gram of U^{235}). The neutrons are slowed down by a moderator, and used to bombard the U^{235} nucleus again, setting up a controlled chain reaction.

The nuclear reactors based on fission using only U^{235} (only about 0.7% of the natural uranium) as the fissile material are called as burner reactors. These reactors waste the abundant U^{238} isotope present in the natural uranium. Although U^{238} is not a fissile material, it is a fertile material which can be converted by neutron bombardment from the fissile material, plutonium 239. Similarly, naturally occurring thorium 232 is also a fertile material. It can be converted into U^{233} which is a fissile material.

The neutrons generated by fission reaction serve two purposes - (a) converting a fertile material into a fissile material and (b) sustain the fission reaction for the fissile material. These reactions are called as breeder reactions if they produce more fissile material than they consume. The nuclear reactor based on these reactions is called breeder reactor.

For India, the Th^{232} to U^{233} breeder cycle is particularly significant as there are large deposits of thorium in the monazite beach sands of Kerala.

Types of Nuclear Reactors

There are five important reactor types used for power generation now.

1. *Boiling Water Reactor (BWR)*: Here steam is raised in the core and flows directly to the turbine.

2. *Pressurized Water Reactor (PWR)*: In PWR, the heated water flows to a heat exchanger where water is converted into steam driving the turbine. The pressure used is greater than 100 atm (1470 lb/in^2) in order the water may not become steam in the pipes.

   The fuel used in BWR and PWR is enriched uranium containing 2.3% of U^{235}. Ordinary water used in both PWR and BWR acts as a moderator and coolant and does not become radioactive except for some atoms of oxygen (O^{18}) present.
3. *Canadian Deuterium Uranium (CANDU)*: Deuterium (D_2O) or heavy water is used both as moderator and coolant in a closed circuit. In the heat exchanger, ordinary water is used to produce steam. The heavy water is under pressure to retain it in liquid state. Natural uranium is used as a fuel.
4. *Gas Cooled Reactor (GCR)*: Here graphite is used as moderator and helium gas as coolant. The heat exchanger is used to heat either water (to produce steam) or helium to run a steam or gas turbine. Natural or enriched uranium is used as fuel.
5. *Liquid Metal Fast Breeder Reactor (LMFBR)*: Here the core containing either fissile U^{235} or Pu^{239} is surrounded by a blanket of fertile U^{238}. There are two liquid sodium loops. The primary loop has the liquid metal as a coolant, flowing in the core and blanket, while the secondary sodium loop takes the heat from the primary sodium loop in the first heat exchanger. It then transfers its heat in a second heat exchanger using ordinary water to raise steam. Since water and sodium form an explosive mixture, water is kept clear of the core and the primary sodium loop.

The first two types of reactors are used primarily in U.S. The third type, CANDU reactor is used in Canada and India. The GCR is used mostly in U.K. and Europe. The LMFBR is used mainly in France. The experimental fast breeder reactor using thorium is under development in India.

Nuclear Fusion

In nuclear fusion, energy is released by joining very light atoms. If the present research on controlled fusion is eventually successful, and fusion reactors are built, they could provide ultimate solution to the

World's energy problem. There are two fusion reactions that are considered most promising: (a) the DT reaction, which fuses deuterium with tritium, making helium and (b) the DD reaction, which fuses deuterium with deuterium.

Deuterium, a naturally occurring heavy isotope of hydrogen can be obtained from sea water. Tritium, a heavier isotope of hydrogen is not found in large quantities naturally, but it can be manufactured from lithium.

A fusion reaction can be self-sustaining only if the rate at which energy is lost from the reacting fuel is less than the rate at which energy is generated by fusion reaction. This is possible only above a certain critical temperature which is estimated to be of the order of 100 million ^{0}C. Thus the key problems in the development of a fusion reactor are the attainment of the required high temperature by initially heating the fuel charge and the confinement of the heated fuel for a long enough time for the reaction to become self-sustaining. The research to solve these problems involves either magnetic confinement or laser induced fusion.

ITER Project

This is an international project that will use DT reaction. It is situated in southern France and is designed to generate 500 MW. The project brings together the scientific and political weight of governments representing more than half the world's population - including the European Union together with China, India, Japan, Russia, South Korea and United States. The overall budget is 15 billion Euros.

According to a news report in August 2013, the project has reached a critical phase. Under the initial plan, it was hoped to achieve the first plasma by the middle of the last decade. Then after a redesign, a new deadline of November 2020 was set. It is now believed that the deadline could be met as the components of the reactor have started arriving.

Since the 1950s, fusion has offered the dream of almost limitless energy - copying the process that powers the sun - fuelled by two readily available forms of hydrogen. The attraction is a combination of cheap fuel, relatively little radioactive waste and no emission of greenhouse gases. However, the technical challenges are immense. The Iter reactor is known as a "tokomak", it is based on Jet, a

European pilot project at Culham in Oxfordshire. It will involve creating a plasma of superheated gas reaching temperature more than 200 million ^{0}C - conditions hot enough to force deuterium and tritium atoms to fuse together and release energy. The whole process will take place inside a giant magnetic field in the shape of a ring - the only way such extreme heat can be contained. The plant at Jet has managed to achieve fusion reactions in very short bursts but required the use of more power than it was able to produce. The reactor at Iter is designed to generate 10 times more power than it will consume.

Discussion

If fusion reactors can be built, it will revolutionize power generation. However, since it is a long way off, we have to depend upon nuclear reactors based on fission. As of July 2013, 30 countries worldwide are operating 434 nuclear reactors and 71 new plants are under construction in 14 countries. The installed capacities in U.S, France and Japan are 102,136 (19%), 63130 (74.8%) and 44,215 (18.1%) MW respectively. In India, the installed capacity is 4780 MW which is only 3.6% of the total installed capacity. Globally, nuclear plants provided 12.3% of electricity in 2011. The contribution of nuclear power generation exceeded 25% in 13 countries.

There are two major issues concerning Nuclear reactors using fission - 1. safety and 2. disposal of radioactive waste. According to David McKay, "Nuclear power is not infinitely dangerous. It is dangerous, much as coal mines, petrol repositories, fossil-fuel burning and wind turbines are dangerous. I think the right way to assess nuclear is to compare it objectively with other sources of power. Coal power stations for example, expose the public to nuclear radiation, because coal ash typically contains uranium". He also proposes a new unit to assess public risk - number of deaths per GWy (Gigawatt year). According to EU figures, coal, lignite and oil have the highest death rates followed by biomass above 1 per GWy. Nuclear and wind are the best with death rates below 0.2 per GWy.

Waste Disposal

The volume of waste from nuclear reactors is relatively small. Most of this waste is low level waste. 7% is intermediate level waste and just 3% is the high level waste. It is conventional to keep the high level waste at the reactor for its first 40 years. It is stored in pools of water and cooled. After 40 years, the radioactivity has dropped 1000 fold. The radioactivity continues to fall and after 1000 years, the level is comparable to that of the uranium ore. Although, 1000 years appear very long period, total high level wastes will occupy the volume of 35 Olympic swimming pools for waste generated in U.K.

C.3 Recent Developments

A recent study by American Physical Society (APS), reported in IEEE Spectrum, December, 2013) states "there are no technical barriers to running some nuclear plants for up to 80 years". The study advocates keeping many of the approximately 100 reactors in the U.S. running for at least 60 years. It also argues that tradeoffs between nuclear and gas is not necessarily a bargain. Nuclear power provides about 100 GW of power, about 20 % of U.S. electricity supply. If its contribution starts to wane after 2030 as reactors close, there could be energy shortfall. According to the author of the APS report, "Nuclear power plants provide the nation with a source of clean energy at a time when renewables such as solar and wind are not yet ready to fill the potential gap in the nation's base power needs created by loss of nuclear power". The nuclear Regulatory Commision (NRC) has already issued extensions to many nuclear plants, raising their retirement age to 60 years, and is evaluating the possibility of some plants being open for 80 years. But it is unclear whether utilities will want to spend the money required to keep the assets up to date.

Nuclear Reactor Renaissance

The nuclear reactor design is undergoing some major changes. While some changes are essentially adding safety features to the existing design, recent developments involve building small modular reactors that produce less than 300 MW, but can be scaled up. In one of the design, the cooling system relies on water convection alone and does not require pumps. The modules can be largely manufactured off-site and transported by barge, truck or rail reducing construction time. The fuel and steam generator are housed inside a water-submerged steel vessel which has a greater ability to withstand pressure and dissipate heat than the building that houses a traditional PWR. In another design, there is no need for a moderator and the reactor is termed the fast reactor.

APPENDIX D

A Thought Experiment: Edison Vs Graham Bell

There is a popular comparison that underscores the pace of change - or lack thereof - regarding our nation's grid. The story goes like this: If Alexander Graham Bell were somehow transported to the 21st century, he would not begin to recognize the components of modern telephony - cell phones, texting, cell towers, PDAs etc - while Thomas Edison, one of the grid's key early architects, would be totally familiar with the grid.

While this thought experiment speaks volumes about appearances. It is far from the whole story. Edison would be quite familiar with the grid's basic infrastructures and perhaps even an electromechanical connection or two, but he would be just as dazzled as Graham Bell with the technology behind the scenes.

Our century-old power grid is the largest interconnected machine on Earth, so massively complex and inextricably linked to human involvement and endeavour that it has alternately (and appropriately) been called an ecosystem. It consists of more than 9200 electric generating units with more than one million megawatts of generating capacity connected to more than 300,000 miles of transmission lines.

In celebrating the beginning of the 21st century, the National Academy of Engineering set about identifying the single most important engineering achievement of the 20th century. The Academy compiled a list of twenty accomplishments which have affected virtually everyone in the developed world. The internet took thirteenth place on this list, and highways eleventh. Sitting at the top of the list was electrification as made possible by the grid, "the most significant engineering achievement of the 20th century".

Given that the growth of the nation's global economic leadership over the past century has in many ways mirrored the trajectory of the grid's development, this choice is not surprising.

In many ways, the present grid works exceptionally well for what it was designed to do-for example, keeping the costs down. Because electricity has to be used the moment it is generated, the grid represents the ultimate in just-in-time product delivery. Everything must work almost perfectly at all times and does.

But, we all of us-have taken this marvelous machine for granted for too long. As a result, our overburdened grid has begun to fail us more frequently and presents us with substantial risks (U.S. DOE Report 2008).

Postscript

A blog posted in Harvard Business Review (2010) on 'The Smart Grid is Wikinomics on a Macro Scale' states "Treating the (power) grid like an open platform would, for example, allow software developers to build applications just as developers build apps for the iPhone".

(Note: No comment is required!)

APPENDIX E

Glossary

Acronyms

ABT : Availability Based Tariff

AC : Alternating Current

ACSR: Aluminum Core Steel Reinforced

AGC : Automatic Generation Control

AMR: Automatic Meter Reading

ASAI: Average Service Availability Index

AVR: Automatic Voltage Regulator

CAES : Compressed Air Energy Storage

CB : Circuit Breaker

CCGT: Combined Cycle Gas Turbine

CSC: Convertible Static Compensator

DAC: Distribution Automation and Control

DC : Direct Current

DFIG: Doubly Fed Induction Generator

DG: Distributed Generation

DISCOM : Distribution Company

DLC : Double Layer Capacitor

DOE: Department of Energy (U.S.)

DR: Demand Response

DS: Distributed Storage

DSM: Demand Side Management

DSTATCOM : Distribution STATCOM

DVR : Dynamic Voltage Restorer

EES : Electric Energy Storage

EHV : Extra High Voltage

EMF: Electro Motive Force

EPRI: Electric Power Research Institute (U.S.)

ESI: Electricity Supply Industry

FACTS : Flexible AC Transmission System

FES : Flywheel Energy Storage

GTO: Gate Turn Off (thyristor)

HV : High Voltage

ICT: Information and Communication Technology

IEC: International Electrotechnical Commission

IGBT: Insulated Gate Bipolar Transistor

IPFC: Interline Power Flow Controller

KCL : Kirchhoff's Current Law

KVL : Kirchhoff's Voltage Law

LFC : Load Frequency Control

MMF: Magneto Motive Force

OLTC: On Load Tap Changing (transformer)

PHS: Pumped Hydro Storage

PLF: Plant Load Factor

POD: Power Oscillation Damper

PQ : Power Quality

PSS: Power System Stabilizer

RES : Renewable Energy Resources

RFB: Redox Flow battery

ROW: Right of Way

SG : Smart Grid

SIL: Surge Impedance Loading

SMES : Superconducting Magnetic Energy Storage

SPEF: Structure Preserving Energy Function

STATCOM : Static (synchronous) Compensator

TCR: Thyristor Controlled Reactor

TCSC: Thyristor Controlled Series Capacitor

TSC: Thyristor Switched Capacitor

UHV : Ultra High Voltage

UI: Unscheduled Interchange

UPFC : Unified Power Flow Controller

UPQC : Unified Power Quality Conditioner

VIU: Vertically Integrated Utility

VSC: Voltage Source Converter

WAMS: Wide Area Measurement System

Kirchchoff's Current Law (KCL): The algebraic sum of currents entering a node (vertex) in an electric circuit is zero. (Note that the current is negative if it is leaving the node).

Kichchoff's Voltage law (KVL): The algebraic sum of potential drops (or rise) around a mesh (or loop) is zero.

ELECTRIC CURRENT (measured in amperes) is the flow of electric charges, similar to the flow water in a pipe. However, the diameter of the electrical conductor, required to carry thousands of amperes is small compared to that of water pipes. Also, the speed of current flow in a wire approaches that of the speed of light.

ELECTRIC VOLTAGE (measured in Volts) is similar to the water pressure. For a specific length and diameter, higher the pressure, greater is the flow.

POWER also called as active power (or real power in AC systems) is the product of voltage and current in DC circuits, whereas there is a power factor which needs to be multiplied by the product of voltage and current in AC circuits. The power factor varies between zero and one. For a AC generator, the rated power factor is typically between 0.8 and 0.9.

REACTIVE POWER in AC circuits is a measure of the rate of energy flow that keeps oscillating between the generator (or a source of power) and the load at twice the frequency. Reactive power equals the average power that would flow to an inductor or capacitor that would result in the total energy supplied from the source in a half cycle and returned in the other half cycle (of a sine wave).

APPARENT POWER is the vector sum of power (P) and reactive power (Q). The complex power is P + jQ. Apparent power is the magnitude of the complex power.

(If we draw a right angled triangle with two sides representing P and Q, the hypotenuse represents the apparent power)

POWER FACTOR is the ratio of power to apparent power.

Impedance in AC circuits is the ratio of voltage phasor to the current phasor and is expressed as a complex number R +jX. The real part is called as resistance (R) and the imaginary part is the reactance (X). An AC transmission line has series impedance. The voltage drop across the resistance is an indicator of the power loss in the line, whereas the voltage drop across the reactance indicates reactive power loss.

Admittance is the reciprocal of the impedance and is expressed as a complex number G + jB, where G is the conductance and B is the susceptance. If X is positive, B is negative and vice versa. In general, G and B are not the reciprocals of R and X respectively.

References

A. Books

Begamudre R.D. (2000), *Energy Conversion Systems,* New Delhi: New Age.

Bush, Stephen F. (2014), *Smart Grid: Communication-Enabled Intelligence for the Electric Power Grid,* New York: John Wiley.

Feynman R.P., R.B. Leighton and M.Sands, (1964), *The Feynman Lectures on Physics,* New Delhi: Addison-Wesley/Narosa

Glover J.D. and M.S.Sarma (2003), *Power System Analysis and Design, 3rd Edition,* New Delhi:Thomson/Brooks/Cole.

Gonen, T., (1986), *Electric Power Distribution System Engineering,* New York: McGraw Hill.

McKay D.J.C.(2009), *Sustainable Energy-without hot air,* Cambridge: UIT Cambridge.

Padiyar K.R. (2002), *Power System Dynamics-Stability and Control, Second Edition,* Hyderabad: BS Publications.

Padiyar K.R. (2007), *FACTS Controllers in Power Transmission and Distribution,* New Delhi: New Age.

Padiyar K.R. (2010), *HVDC Transmission Systems, Second Edition,* New Delhi: New Age.

Padiyar K.R. (2013), *Structure Preserving Energy Functions in Transmission and Distribution,* Boca Raton: CRC Press.

Slemon G.R. and A.Straughen (1980), *Electric Machines,* Reading: Addison-Wesley.

Sukhatme S.P. (1996), *Solar Energy-Principles of Thermal Collection and Storage,* New Delhi: Tata McGraw-Hill.

Wood A.J. and R.F. Wollenberg (1996), *Power Generation, Operation and Control, Second Edition,* New York: John Wiley.

B. Reports and Articles

Amin, S.M.(2011), U.S.Electrical grid gets less reliable, *IEEE Spectrum*, V.48.

Bhanu Bhushan (2005), *ABC of ABT: A Primer on Availability Tariff,*

Casazza J., F.Delea and G.Loehr (2007), *Electrons versus People, Which Group is Smarter?*, IEEE Power and Energy Magazine, V. 5, No. 1.

Dalziel C.F.(1972), Electric shock hazard, *IEEE Spectrum* pp 41-50

Electric Power Research Institute(EPRI) (2010), *Electricity Energy Storage Technology Options-A White Paper Primer on Applications, Costs and Benefits*

Geddes L.A. and L.E.Baker (1971), Response to passage of electric current through the body, *J. Assoc. Advancement of Medical Instr.*, V.2, pp.13-18.

Hingorani N.G.(1995), Introducing Custom Power, *IEEE Spectrum*, (32):6, pp.41-48.

International Electrotechnical Commission (2010), *IEC Smart Grid Standardization Road Map.*

International Electrotechnical Commission (IEC) (2011), *Electrical Energy Storage-White paper.*

Lerner E.J. (2003), *What's wrong with the Electric Grid ?,* The Industrial Physics, American Institute of Physics.

Mills M.P. (2013) *The Cloud Begins with Coal-Big Data, Big Networks, Big Infrastructure and Big Power- An Overview of the Electricity Used by the Global Digital Ecosystem,* Sponsored by American Coalition for Clean Coal Electricity.

Oberschmidt, M. et al.(2013), Techno-economic analysis of electricity storage systems, Chapter 10, Electricity transmission, distribution and storage systems, Edited by Z.Melhem, Woodhead Publishing, Oxford.

Reider W, (1970), Circuit Breakers: Physical and Engineering Problems, I Fundamentals, *IEEE Spectrum*, V(7): 7, pp.35-43.

Reilly J.F. (1979), Short term effects on people, *IEEE Tutorial course on 'The Electrostatic and Electromagnetic Effects of AC Transmission Lines' (79 EH0145-3-PWR).*

Russell W. et al. (2001), *Restructuring vs Reliability- A Tale of Declining Standards,* American Public Power Association.

Sachs J.D. (2008), *Blackouts and cascading Failures on the Global markets,* Scientific American magazine.

Sally Adee and Enrico Guizzo, (2010), *Nuclear Reactor Renaissance,* IEEE Spectrum, July 2010.

U.S. Department of Energy (2003), *GRID 2030, A National Vision for Electricity's Second 100 Years,* Office of the Electric Transmission and Distribution.

U.S. Department of Energy (2008), *The Smart Grid: An Introduction.*

U.S.-Canada Power System Outage Task Force (2004), *Final Report on the August 14, 2003 Blackout in the United States and Canada: Causes and Recommendations.*

Van Doren P. and J. Taylor (2004), *Rethinking Electricity Restructuring,* Policy Analysis No.530, Cato Institute.

World Health Organization (2006), Electromagnetic fields and public health, *Fact Sheet No. 299, March 2006.*

Index

E

F

G

H

I

L

M

N

O

P

R

S

T

U

V

W

X

Z

9 789385 433870

Printed at Repro India Ltd.